MULTIPLE CHOICE AND

Higher HUMAN BIOLOGY

for CfE

Writing Team:

James Torrance

James Fullarton

Clare Marsh

James Simms

Caroline Stevenson

Diagrams by James Torrance

SCOTTISH EXAMINATION MATERIALS

HODDER GIBSON
AN HACHETTE UK COMPANY

The Publishers would like to thank the following for permission to reproduce copyright material:

Photo credits p.1 (background) and Unit 1 running head image © Alexandr Mitiuc - Fotolia.com; p.1 (inset left) © STEVE GSCHMEISSNER/SCIENCE PHOTO LIBRARY, (inset centre) © 3D4MEDICAL.COM/SCIENCE PHOTO LIBRARY, (inset right) © Michael J. Gregory, Ph.D./Clinton Community College Michael.Gregory@clinton.edu; p.59 (background) and Unit 2 running head image © Imagestate Media (John Foxx) / Patient Care V3063; p.59 (inset left) © Dan Marschka/AP/Press Association Images, (inset centre) © DR NAJEEB LAYYOUS/SCIENCE PHOTO LIBRARY, (inset right) © DR P. MARAZZI/SCIENCE PHOTO LIBRARY; p.105 (background) and Unit 3 running head image © Sebastian Kaulitzki – Fotolia; p.105 (inset left) © BSIP SA / Alamy, (inset centre) © STEVE GSCHMEISSNER/SCIENCE PHOTO LIBRARY, (inset right) © MEDICAL BODY SCANS/JESSICA WILSON/SCIENCE PHOTO LIBRARY; p.143 (background) and Unit 4 running head image © Sebastian Kaulitzki – Fotolia; p.143 (inset left) © DAVID SCHARF/SCIENCE PHOTO LIBRARY, (inset centre) © DAVID SCHARF/SCIENCE PHOTO LIBRARY, (inset right) © JOHN BAVOSI/SCIENCE PHOTO LIBRARY.

Every effort has been made to trace all copyright holders, but if any have been inadvertently overlooked the Publishers will be pleased to make the necessary arrangements at the first opportunity.

Although every effort has been made to ensure that website addresses are correct at time of going to press, Hodder Gibson cannot be held responsible for the content of any website mentioned in this book. It is sometimes possible to find a relocated web page by typing in the address of the home page for a website in the URL window of your browser.

Hachette UK's policy is to use papers that are natural, renewable and recyclable products and made from wood grown in sustainable forests. The logging and manufacturing processes are expected to conform to the environmental regulations of the country of origin.

Orders: please contact Bookpoint Ltd, 130 Park Drive, Milton Park, Abingdon, Oxon OX14 4SE. Telephone: (44) 01235 827720. Fax: (44) 01235 400454. Lines are open 9.00–5.00, Monday to Saturday, with a 24-hour message answering service. Visit our website at www.hoddereducation.co.uk. Hodder Gibson can be contacted direct on: Tel: 0141 848 1609; Fax: 0141 889 6315; email: hoddergibson@hodder.co.uk

© James Torrance, James Fullarton, Clare Marsh, James Simms, Caroline Stevenson 2015

First published in 2015 by

Hodder Gibson, an imprint of Hodder Education,

An Hachette UK Company,

2a Christie Street

Paisley PA1 1NB

Impression number   5 4 3 2 1

Year   2019 2018 2017 2016 2015

Cover photo © V. Yakobchuk – Fotolia.com

Illustrations by James Torrance

Typeset in Minion Regular 11/14 pt by Integra Software Services Pvt. Ltd., Pondicherry, India

Printed in Slovenia

A catalogue record for this title is available from the British Library

ISBN: 978 1 4718 4743 1

Contents

Preface

This book has been written specifically to complement the textbook *Higher Human Biology for CfE*. It is intended to act as a valuable resource for pupils and teachers by providing a set of matching exercises and a comprehensive bank of multiple choice questions, the content of which adheres closely to the SQA syllabus for Higher Human Biology for CfE.

Each test corresponds to a key area of the syllabus and to a chapter in the textbook. The matching exercises enable pupils to gradually construct a glossary of terms essential to the course. The multiple choice components contain a variety of types of item, many testing *knowledge and understanding*, some testing *problem-solving skills* and others testing *practical abilities*. These allow pupils to practise extensively in preparation for the examination. The book concludes with two 20-item specimen examinations in the style of the multiple choice section of the externally assessed higher examination paper.

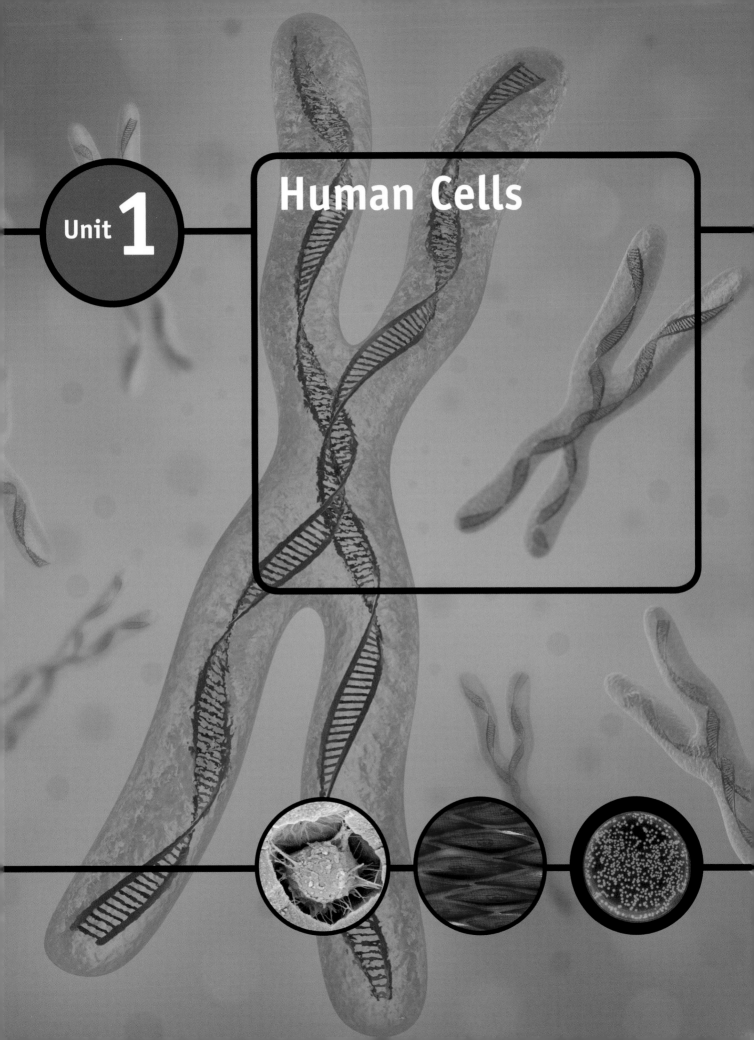

Unit **1**

Human Cells

1 Division and differentiation in human cells

Matching Test

Match the terms in list X with their descriptions in list Y.

list X

1 blastocyst
2 bone marrow
3 cancer
4 differentiation
5 ethics
6 germline
7 meiosis
8 mitosis
9 model cells
10 multipotent
11 pluripotent
12 regulation
13 somatic
14 specialised
15 therapeutic

list Y

a) division of a diploid germline cell's nucleus and cytoplasm to form four haploid gametes

b) division of a diploid nucleus into two diploid nuclei, followed by cell division to form two identical daughter cells

c) form of control to ensure the quality of stem cells used and the safety of the procedures carried out

d) type of diploid cell in humans that divides by meiosis to form haploid gametes

e) type of diploid cell in humans that divides by mitosis (but never by meiosis) and forms more diploid cells

f) moral values that ought to govern human conduct

g) common source of tissue (adult) stem cells in humans

h) early human embryo composed of unspecialised cells

i) term describing the use of stem cells in the repair of damaged or diseased organs

j) cell that has become differentiated and only expresses genes for proteins specific to that cell type

k) process of cell specialisation involving the selective switching off and on of certain genes

l) type of stem cell capable of differentiating into a limited range of cell types

m) name given to stem cells used to test drugs and study how diseases develop

n) type of stem cell capable of differentiating into almost all of a human being's cell types

o) uncontrolled growth of cells leading to the production of a tumour

Multiple Choice Test

Choose the ONE correct answer to each of the following multiple choice questions.

Questions 1 and 2 refer to Figure 1.1, which represents the processes of cell division and cellular differentiation in a human.

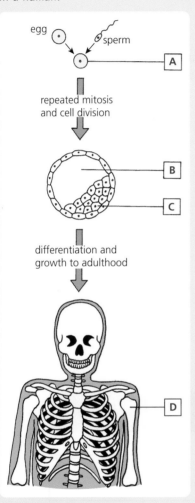

Figure 1.1

1 Which box indicates the location of a tissue (adult) stem cell?
2 Which box indicates a cell type that, on being cultured in a laboratory, would give rise to embryonic stem cells?

Questions 3 and 4 refer to the following possible answers.

    A multipotent   **B** totipotent   **C** pluripotent   **D** omnipotent

3 Which term correctly refers to embryonic stem cells?
4 Which term correctly refers to tissue (adult) stem cells?
5 Which of the following statements about stem cells is NOT true?
    A They are unspecialised cells present in multicellular animals.
    B They are able to reproduce themselves by repeated mitosis and cell division.
    C They are multipotent cells present in the stems of all green plants.
    D They differentiate into specialised cells when required to do so.

Questions 6 and 7 refer to the following possible answers.

   A connective   **B** epithelial   **C** muscle   **D** nerve

6 Which of these types of body tissue includes bone and cartilage?

7 Which of these types of body tissue acts as a covering that lines the body cavities and tubular structures?

8 Which of the following cell types are haematopoietic stem cells UNABLE to make?

   A phagocytes   **B** germline cells   **C** lymphocytes   **D** red blood cells

9 In humans, each somatic cell is

   A haploid and contains 23 chromosomes.   **B** haploid and contains 23 pairs of chromosomes.

   C diploid and contains 23 chromosomes.   **D** diploid and contains 23 pairs of chromosomes.

10 Which letter in Figure 1.2 of the human life cycle represents a germline cell?

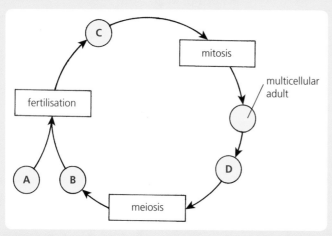

Figure 1.2

11 Which row in Table 1.1 correctly refers to a germline cell in a human being?

	Ploidy state of cell	Products of nuclear and cell division(s)	Type of nuclear division involved
A	haploid	2 identical daughter cells	meiosis
B	diploid	4 haploid gametes	meiosis
C	haploid	2 identical daughter cells	mitosis
D	diploid	4 haploid gametes	mitosis

Table 1.1

12 The five boxes in Figure 1.3 describe stages in the procedure carried out during bone marrow transplantation.

P

Freezing

bone marrow or blood frozen to preserve stem cells until patient has completed chemotherapy

Q

Collection

stem cells collected from donor's bone marrow or peripheral blood

R

Infusion

thawed stem cells infused into patient to engraft in bone marrow and make normal blood cells

S

Processing

bone marrow or peripheral blood processed to concentrate stem cells

T

Chemotherapy

patient given high dose of chemotherapy and/or radiation to destroy cancerous cells in bone marrow

Figure 1.3

Their correct sequence is

A Q, P, S, T, R.   **B** S, Q, T, R, P.   **C** Q, S, P, T, R.   **D** S, Q, P, R, T.

13 Under natural conditions, a tissue (adult) stem cell of a human being has
   A many of its genes switched off and is only capable of giving rise to a limited range of cell types.
   B many of its genes switched off but is capable of differentiating into any cell type.
   C all of its genes switched on but is only capable of giving rise to a limited range of cell types.
   D all of its genes switched on and is capable of differentiation into any cell type.

14 Table 1.2 shows the result of an experiment set up to investigate the effect of a growth-stimulating factor on a type of stem cell. Flask 1 contained nutrient-rich medium and growth-stimulating factor. Flask 2 contained nutrient-rich medium only. Which of the graphs in Figure 1.4 represents the results correctly?

| Time (days) | Mean cell number × 10⁵/cm³ | |
	flask 1	flask 2
0	3	3
2	4	5
4	9	6
6	11	3
8	15	2
10	24	1

Table 1.2

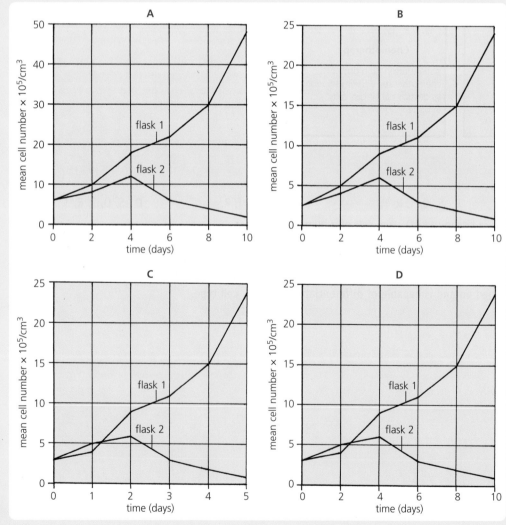

Figure 1.4

Questions 15 and 16 refer to Figure 1.5, which shows a simplified version of the means by which 'Dolly' the sheep was produced.

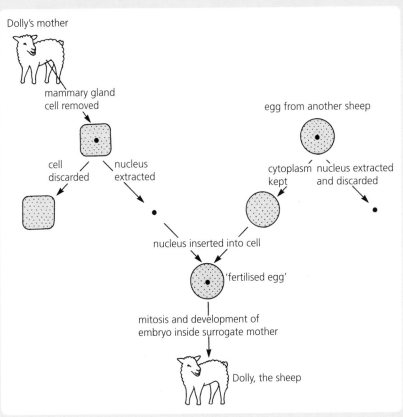

Figure 1.5

15 The name of the process employed to create the original cell that gave rise to Dolly is called
    A nuclear transfer technique.   B multipotent cell-culturing.
    C transplantation of embryonic cell.   D induction of pluripotent stem cell.

16 Which of the following statements is FALSE?
    A Dolly and her mother contained genetically identical nuclear material.
    B A mammary gland cell in a sheep contains repressed genes that can become switched on again.
    C Differentiation of the mammary gland cells in a sheep is an irreversible process.
    D During Dolly's embryonic development, cells became differentiated and specialised.

17 The use of stem cells raises several ethical issues. Ethics refers to the
    A moral values that ought to govern human conduct.
    B techniques followed when a procedure is closely regulated.
    C safety standards that must be maintained during research work.
    D qualifications of the experts employed to carry out an investigation.

Questions 18 and 19 refer to Table 1.3, which shows data referring to bowel cancer deaths in the UK during 2010–2012.

Age group (years)	Average number of deaths per year		Mortality rate (deaths per 100 000 population per year)	
	male	female	male	female
0–9	0	0	0.0	0.0
10–19	1	0	0.0	0.0
20–29	13	17	0.3	0.4
30–39	56	48	1.4	1.2
40–49	219	188	4.8	4.0
50–59	728	500	19.4	12.9
60–69	1820	1100	55.6	32.1
70–79	2775	1850	137.0	78.2
80+	3020	3518	288.5	190.2

Table 1.3

18 In males, the average number of deaths among age group 60–69 is greater than that among age group 20–29 by a factor of

A 65 times.   B 140 times.   C 1083 times.   D 1807 times.

19 Which of the following conclusions can be correctly drawn from the data?
   A Within every age group, more men than women died of bowel cancer annually.
   B The average number of female deaths increased by 150% from age group 50–59 to 60–69.
   C The greatest difference in average number of deaths between males and females occurred in group 60–69.
   D At age 80+ years, the mortality rate for women was less than that for men.

20 The following bar graph shows prostate cancer mortality rates per 100 000 population in the UK in 2012. (The error bars indicate a 95% level of confidence.)

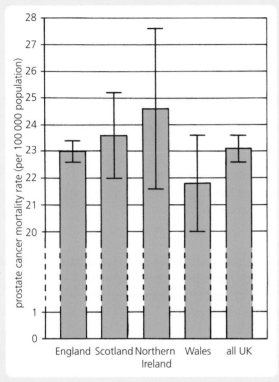

Figure 1.6

Which of the following conclusions about the mortality rate for prostate cancer can be correctly drawn from these results (with a 95% level of confidence)?

A It is significantly lower for Wales than for Scotland.

B It does not differ significantly between countries in the UK.

C It is significantly higher for Northern Ireland than for England.

D It is found in Wales to be significantly lower than the all-UK average.

2 Structure and replication of DNA

Matching Test Part 1
Match the terms in list X with their descriptions in list Y.

list X
1 3' end
2 5' end
3 adenine (A)
4 antiparallel
5 cytosine (C)
6 deoxyribose
7 DNA
8 double helix
9 genotype
10 guanine (G)
11 nucleotide
12 sugar–phosphate backbone
13 thymine (T)

list Y
a) basic unit from which nucleic acids are composed
b) term used to describe two strands of DNA with their sugar–phosphate backbones running in opposite directions
c) supporting structure of nucleic acid molecule formed by bonding between adjacent nucleotides
d) sugar present in DNA
e) base present in DNA that is complementary to adenine
f) base present in DNA that is complementary to guanine
g) base present in DNA that is complementary to cytosine
h) base present in DNA that is complementary to thymine
i) nucleic acid present in chromosomes
j) genetic constitution of an organism determined by the sequence of bases in its DNA
k) phosphate end of DNA strand to which nucleotides cannot be added
l) deoxyribose end of DNA strand to which nucleotides can be added
m) two-stranded molecule of DNA wound into a spiral

Matching Test Part 2
Match the terms in list X with their descriptions in list Y.

list X
1 complementary
2 DNA
3 DNA polymerase
4 ligase
5 primer
6 replication
7 template strand

list Y
a) term used to refer to the unwound strand of DNA to be replicated
b) term used to describe two members of a base pair able to join by hydrogen bonding
c) complex molecule present in chromosomes that stores genetic information
d) process by which a molecule of DNA reproduces itself
e) enzyme required to promote DNA replication
f) enzyme that joins replicated DNA fragments into a complete strand
g) short sequence of nucleotides needed by DNA polymerase to begin replication of DNA

Multiple Choice Test

Choose the ONE correct answer to each of the following multiple choice questions.

1 The structure of one nucleotide is shown in Figure 2.1.

Which of the diagrams in Figure 2.2 shows two nucleotides correctly joined together?

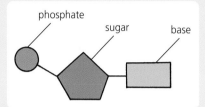

Figure 2.1

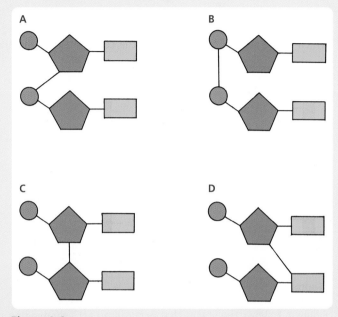

Figure 2.2

2 If a double-stranded DNA molecule is 50 000 base pairs long, how many nucleotides does it contain?

  A 25 000   **B** 50 000   **C** 100 000   **D** 200 000

3 A shorthand method of representing part of a single strand of DNA is shown in Figure 2.3.

Which part of Figure 2.4 shows the correct positions of the phosphate (P), sugar (S) and base (B) molecules?

Figure 2.3

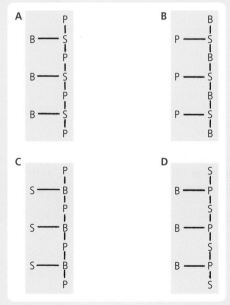

Figure 2.4

4   If a DNA molecule contains 10 000 base molecules, of which 18% are thymine, then the number of cytosine molecules present is

    A 1800   **B** 3200   **C** 6400   **D** 8200

5   If a DNA molecule contains 4000 base molecules and 1200 of these are adenine, then the percentage number of guanine bases present in the molecule is

    A 12   **B** 20   **C** 28   **D** 30

6   Figure 2.5 shows part of one strand of a DNA molecule.

Which of the strands in Figure 2.6 is the complement of the original strand?

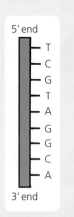 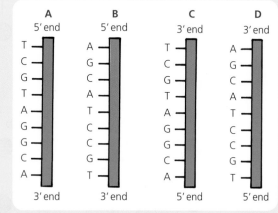

Figure 2.5   **Figure 2.6**

7   Figure 2.7 shows the stages that occur in an actively dividing mammalian cell.

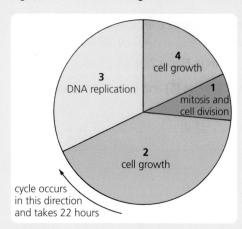

Figure 2.7

The drug aminopterin inhibits thymine production. When this drug is added to a culture of actively dividing cells, after 16 hours most of the cells are found to have been UNABLE to complete one of the stages in the cycle. This stage is number

    A 1   **B** 2   **C** 3   **D** 4

8   Figure 2.8 represents Griffith's bacterial transformation experiment.

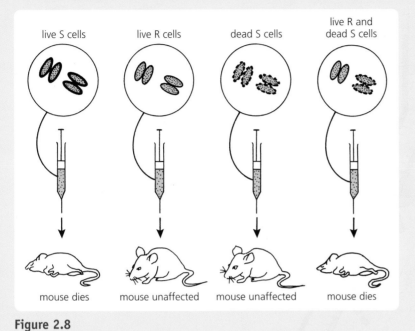

Figure 2.8

Which syringe in Figure 2.9 would NOT lead to the death of a mouse?

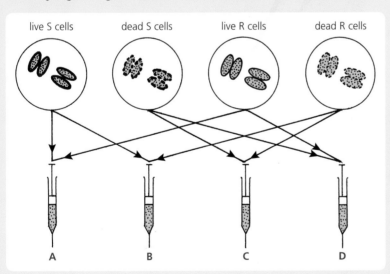

Figure 2.9

9   Applying Chargaff's rules relating to the number of different bases in a DNA sample, which of the following
     is correct?

A   T = C   B   G = A   C   C + G = T + A   D   T + C = G + A   ➡

Questions 10, 11 and 12 refer to Figure 2.10, which shows the formation of the lagging strand of DNA during replication.

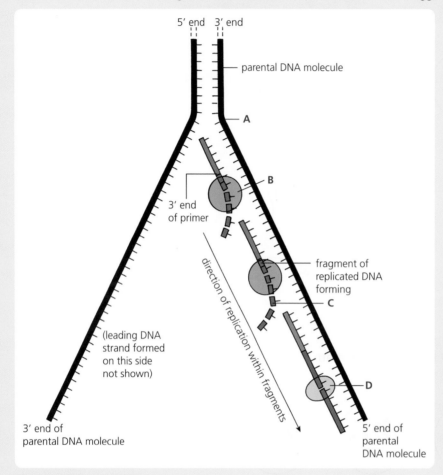

Figure 2.10

10 Which structure is ligase?
11 Which structure is DNA polymerase?
12 Which structure is part of a replication fork?

13 Figure 2.11 shows a portion of DNA undergoing semi-conservative replication.

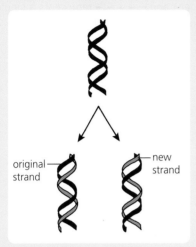

original strand — new strand

Figure 2.11

If the products undergo a further round of semi-conservative replication, which of the sets of products shown in Figure 2.12 would be the result?

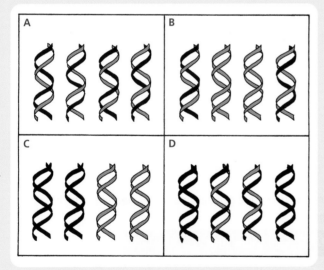

Figure 2.12

Questions 14 and 15 refer to the information shown in Figure 2.13, which shows three types of DNA molecule. ———— represents a single DNA strand labelled with ^{15}N (a heavy isotope of nitrogen) and – – – – – represents a single DNA strand labelled with ^{14}N (the common isotope of nitrogen). A mixture of these can be separated as indicated.

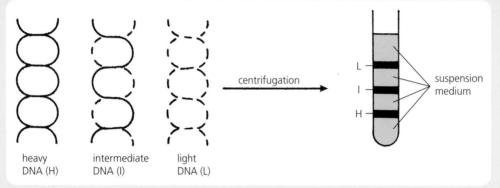

Figure 2.13

14 If bacteria grown for many generations in ^{15}N are transferred to ^{14}N medium for one generation's growth, which of the following will result on extracting and centrifuging their DNA?

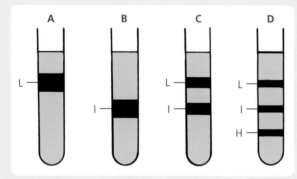

Figure 2.14

15 Imagine that a molecule of heavy DNA is replicated for two successive generations in ^{14}N medium. Which row in Table 2.1 shows the DNA molecules that would be formed?

	Heavy DNA	Intermediate DNA	Light DNA
A	0	100%	0
B	50%	0	50%
C	0	50%	50%
D	50%	50%	0

Table 2.1

3 Gene expression

Matching Test Part 1
Match the terms in list X with their descriptions in list Y.

list X
1 exon
2 genotype
3 intron
4 phenotype
5 primary transcript
6 ribose
7 RNA polymerase
8 splicing
9 transcription
10 uracil

list Y
a) process by which a complementary molecule of mRNA is made from a region of a DNA template
b) base present in RNA that is complementary to adenine
c) sugar present in RNA
d) enzyme that controls transcription
e) coding region of a gene
f) non-coding region of a gene
g) joining of exons from a primary transcript of RNA following the removal of introns
h) mRNA strand formed as the complement of a DNA template strand
i) genetic constitution of a cell or organism determined by the sequence of bases in its DNA
j) physical and chemical state of a cell or organism determined by proteins produced as a result of gene expression

Matching Test Part 2
Match the terms in list X with their descriptions in list Y.

list X
1 amino acid
2 anticodon
3 attachment site
4 codon
5 genetic code
6 mature transcript of mRNA
7 polypeptide
8 post-translational modification
9 ribosome
10 translation
11 tRNA

list Y
a) alteration of protein molecule by cutting and combining its polypeptides
b) molecular language made up of 64 codewords
c) conversion of the genetic information on mRNA into a sequence of amino acids in a polypeptide
d) type of nucleic acid that carries a specific amino acid to a ribosome
e) type of nucleic acid that carries a copy of the DNA code from the nucleus to a ribosome
f) one of 20 different types of organic compound that are the basic building blocks of proteins
g) sub-cellular structure made of rRNA and protein that is the site of protein synthesis
h) triplet of bases on a tRNA molecule that is complementary to an mRNA codon
i) unit of genetic information consisting of three mRNA bases
j) region on a tRNA molecule to which a specific amino acid becomes temporarily fixed
k) chain-like molecule composed of several amino acids

Multiple Choice Test
Choose the ONE correct answer to each of the following multiple choice questions.

1 One of the nucleotides present in mRNA has the composition
  A adenine – ribose – phosphate.
  B uracil – deoxyribose – phosphate.
  C thymine – ribose – phosphate.
  D guanine – deoxyribose – phosphate.

➡

2 Which row in Table 3.1 is correct?

	Present in DNA	**Present in RNA**
A	uracil	thymine
B	ribose	deoxyribose
C	double strand	single strand
D	four different nucleotides	five different nucleotides

Table 3.1

3 Strand X in Figure 3.1 shows a small part of a nucleic acid molecule.

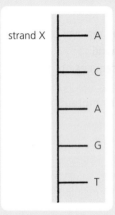

strand X

— A

— C

— A

— G

— T

Figure 3.1

Which pair of strands in Figure 3.2 are complementary to strand X?

A 1 and 2   **B** 2 and 4   **C** 1 and 3   **D** 3 and 4

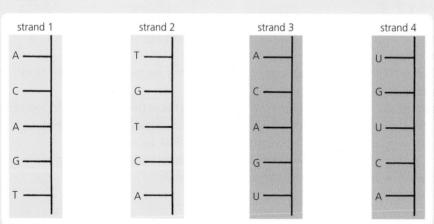

Figure 3.2

Questions 4 and 5 refer to Figure 3.3, which represents transcription of a small portion of mRNA from a template strand of DNA.

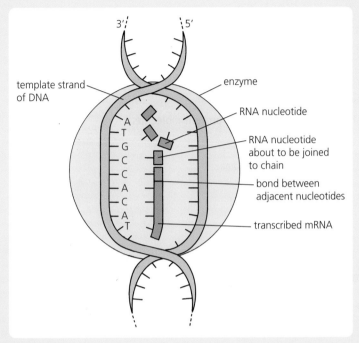

Figure 3.3

4 The enzyme responsible for this process is
  A ligase.
  C DNA polymerase.
  B polypeptidase.
  D RNA polymerase.

5 Which part of Figure 3.4 correctly shows the transcribed mRNA?

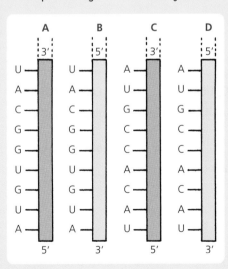

Figure 3.4

6 A free transfer RNA molecule can combine with
  A one specific amino acid only.
  C three different amino acids.
  B any available amino acid.
  D a chain of amino acids.

Questions 7, 8, 9 and 10 refer to the following possible answers:

  A DNA   **B** tRNA   **C** mRNA   **D** amino acid

  7 On which type of molecule are anticodons found?

  8 Which of these molecules bears codons that are complementary to anticodons?

  9 Which of these molecular types must be present in the largest number for successful synthesis of a large protein molecule to occur?

10 Which of these types of molecule holds the master copy of the genetic information that determines the order in which amino acids are joined into a growing protein chain?

Questions 11 and 12 refer to Figure 3.5, which shows the synthesis of part of a protein molecule.

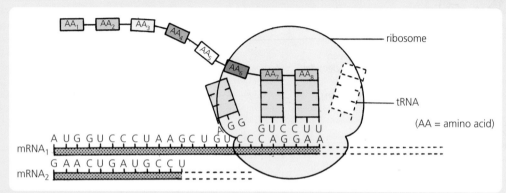

Figure 3.5

11 Which of the following is the first part of the protein molecule that would be translated from mRNA$_2$?

  A AA$_4$ – AA$_2$ – AA$_7$ – AA$_6$ **B** AA$_6$ – AA$_7$ – AA$_2$ – AA$_4$

  C AA$_3$ – AA$_1$ – AA$_5$ – AA$_8$ **D** AA$_8$ – AA$_5$ – AA$_1$ – AA$_3$

12 Figure 3.6 shows a small part of a different protein that was also synthesised on this ribosome.

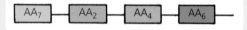

Figure 3.6

What sequence of bases in DNA coded for this sequence of amino acids?

  A CAGGUCAAGUCC   **B** GTCCAGTTCAGG   **C** GUCCAGUUCAGG   **D** CAGCTCAAGTCC

13 If each amino acid weighs 100 mass units, what is the weight (in mass units) of the protein molecule synthesised from an mRNA molecule that is 600 bases long?

  A 2000   **B** 6000   **C** 20 000   **D** 60 000

14 A variety of different proteins can be expressed from the same gene as a result of

  A alternative splicing of RNA and pretranslational modification.

  B alternative splicing of DNA and pretranslational modification.

  C alternative splicing of RNA and post-translational modification.

  D alternative splicing of DNA and post-translational modification.

15 When a cell's DNA becomes damaged, phosphate is added to a regulatory protein called p53, making it become active p53 tumour-suppressor protein. This process is an example of

  A cleavage of exons.   **B** alternative RNA splicing.

  C transcription of introns.   **D** post-translational modification.

4 Genes and proteins in health and disease

Matching Test Part 1

Match the terms in list X with their descriptions in list Y.

list X
1 amino acid
2 antibody
3 enzyme
4 gel electrophoresis
5 hormone
6 hydrogen bond
7 nitrogen
8 peptide bond
9 polypeptide
10 protein

list Y
a) chemical element present in all proteins in addition to carbon, hydrogen and oxygen
b) weak chemical link holding a polypeptide chain in a coil
c) type of chemical (often protein-based) that acts as a chemical messenger
d) type of protein made by white blood cells to defend the body against antigens
e) chain-like molecule composed of several amino acids
f) molecule composed of one or more polypeptides folded or coiled into a specific shape
g) strong chemical link joining adjacent amino acids in a polypeptide chain
h) type of protein possessing an active surface that combines with a specific substrate
i) one of 20 different types of organic compound that are the basic building blocks of proteins
j) technique used to separate electrically charged molecules by subjecting them to an electric current that forces them to move through a sheet of gel

Multiple Choice Test Part 1

Choose the ONE correct answer to each of the following multiple choice questions.

1 The number of different types of amino acid commonly found to make up proteins is
   A 20   B 64   C 200   D 640

2 The connections that join individual amino acid molecules into a long chain are called
   A hydrogen bonds.   B protein bonds.
   C peptide bonds.   D sugar-phosphate bonds.

3 Which of the following chemical elements is always a constituent of protein?
   A calcium   B iron   C nitrogen   D sodium ➡

4 During the process of gel electrophoresis, electrically charged molecules are subjected to an electric current that forces them to move through a sheet of gel. The smaller the molecule, the further the distance it travels. Figure 4.1 shows four stages employed during the separation of DNA by this process. The order in which they would be carried out is

A 4,2,1,3   B 2,4,3,1   C 2,4,1,3   D 4,2,3,1

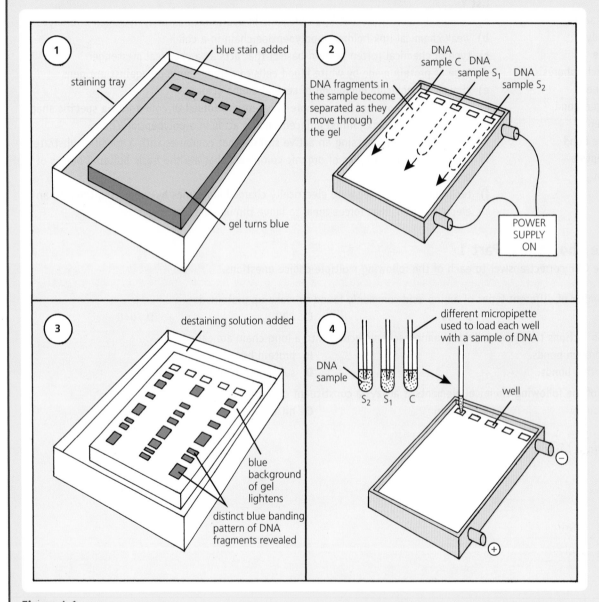

Figure 4.1

Questions 5, 6 and 7 refer to Figure 4.2, which shows the result of employing gel electrophoresis to produce protein 'fingerprints' of five samples taken from fish P, Q, R, S and T.

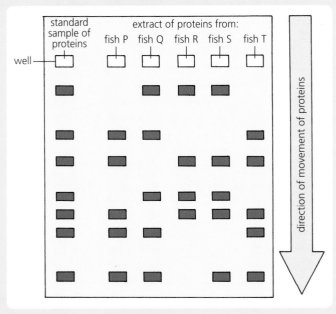

Figure 4.2

5 Which of the following fish protein extracts differs in THREE ways from the standard sample of fish proteins?
A P   B Q   C R   D S

6 Which of the following pairs of fish protein extracts differ from one another in FOUR ways?
A P and R   B P and S   C Q and R   D R and T

7 Which two protein extracts are most likely to belong to the same species of fish?
A P and T   B R and S   C Q and R   D P and Q

8 Table 4.1 gives the mass per 100 g of protein of five different amino acids found in FOUR proteins.

	Protein	Mass of amino acid (g/100 g protein)				
		glycine	alanine	leucine	valine	phenylalanine
A	insulin	4	5	13	8	8
B	haemoglobin	6	7	15	9	8
C	keratin	7	4	11	5	4
D	albumin	3	7	9	7	8

Table 4.1

Which protein is represented by the following pie chart?

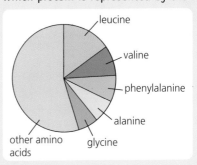

Figure 4.3

23

9 Figure 4.4 shows the sequence of amino acids present in one molecule of insulin. In this protein the ratio of leucine : glycine : tyrosine : histidine is

A 6: 4: 3: 2   B 6: 4: 4: 1   C 3: 2: 1: 1   D 3: 2: 2: 1

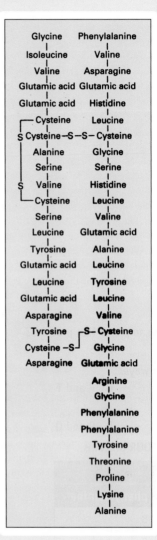

Figure 4.4

10 Choose the one correct pair of answers needed to complete the sentence:

_____ are to proteins as nucleotides are to _____.

A amino acids; ribosomes   C enzymes; ribosomes

B enzymes; nucleic acids   D amino acids; nucleic acids

Matching Test Part 2

Match the terms in list X with their descriptions in list Y.

list X
1 deletion
2 duplication
3 frame-shift
4 gene
5 insertion
6 missense
7 mutant
8 mutation
9 nonsense
10 splice-site mutation
11 substitution
12 translocation

list Y
a) sequence of DNA bases that codes for a protein
b) general term for a random change in an organism's genome
c) individual whose genotype expresses a mutation
d) change affecting nucleotide(s) at a site where introns are normally removed from a primary mRNA transcript
e) general term for a type of point mutation that affects the triplet grouping, thereby altering every subsequent codon along the gene's DNA strand
f) gene mutation involving the addition of an extra nucleotide to the DNA sequence
g) gene mutation involving the exchange of one nucleotide for another in the DNA sequence
h) gene mutation involving the loss of one nucleotide from the DNA sequence *or* chromosome structure mutation involving loss of one or more genes
i) chromosome structure mutation involving the transfer of a segment of genes from one chromosome to another non-matching chromosome
j) chromosome structure mutation where part of a chromosome involving several genes becomes doubled
k) type of substitution where the altered codon for an amino acid makes sense but not the original sense
l) type of substitution where the altered codon acts as a premature stop codon and halts protein synthesis

Multiple Choice Test Part 2

Choose the ONE correct answer to each of the following multiple choice questions.

1 A mutation is a
   A gradual change in hereditary material directed by a changing environment.
   B change in phenotype followed by a change in genotype.
   C sudden temporary change in an organism's genetic material.
   D change in genotype that may result in a faulty protein being expressed.

2 Figure 4.5 shows four of the steps involved in the streak method of isolating yeast from grape 'juice' containing yeast bloom on a Petri dish of agar containing selective medium.

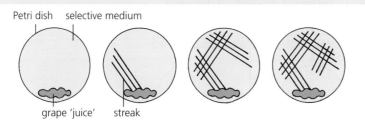

Figure 4.5

Which part of Figure 4.6 shows the missing step in the streaking procedure?

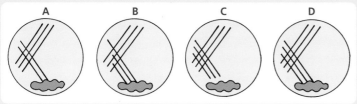

Figure 4.6

3   Figure 4.7 shows the serial dilution of a colony of yeast cells about to be carried out.

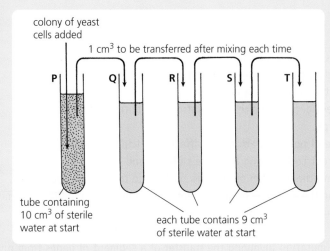

Figure 4.7

Once the procedure has been completed, by how many times will the concentration of yeast cells in tube Q exceed that in tube T?

A 10^2 B 10^3 C 10^4 D 10^5

Questions 4 and 5 refer to Figure 4.8, which shows the results of an investigation using UV-sensitive yeast to test the effectiveness of sun barrier creams.

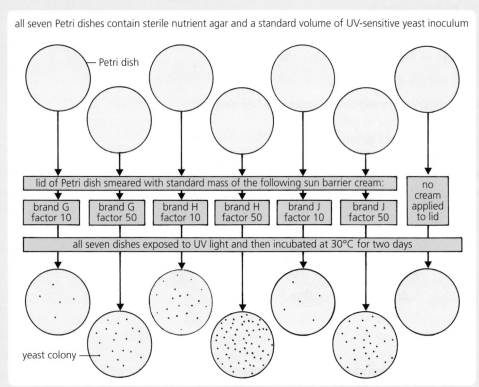

Figure 4.8

4 From the results it can be concluded that
   A brand G offers more protection against UV light than brand H at factor 50.
   B brand G offers more protection against UV light than brand J at factor 10.
   C brands H and J offer equal protection against UV light at factor 50.
   D brands G and J offer less protection against UV light than brand H at factor 10.

5 A further control that could have been included is a Petri dish containing sterile
   A nutrient agar and UV-sensitive yeast, kept in darkness.
   B nutrient agar and wild type yeast, given UV light.
   C plain agar and UV-sensitive yeast, given UV light.
   D plain agar and wild type yeast, kept in darkness.

6 Table 4.2 shows the results from an investigation into the effect of increased dosage of X-rays on frequency of lethal mutations in fruit flies.

Dosage of X-rays (roentgens)	Frequency of lethal mutations (%)
1000	3
2000	5
4000	13
8000	23

Table 4.2

Which of the following graphs correctly represents the results as a line of best fit?

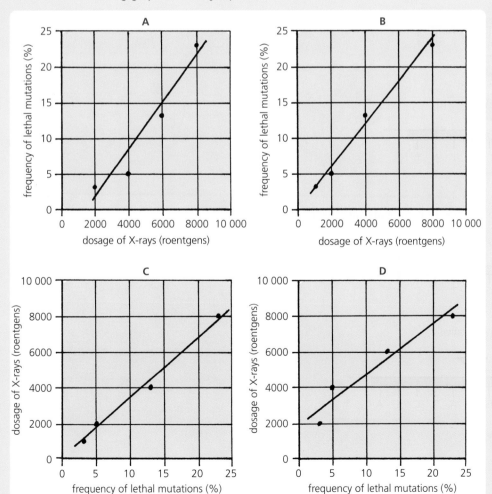

Figure 4.9

7 Which of the following is NOT a mutagenic agent?

  A low temperature **B** gamma rays **C** mustard gas **D** ultraviolet rays

Questions 8, 9 and 10 refer to the following possible answers.

  A translocation **B** deletion **C** insertion **D** substitution

8 What name is given to the type of point mutation illustrated in Figure 4.10?

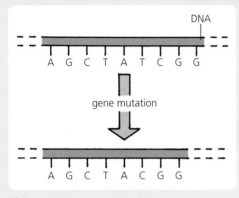

Figure 4.10

9 What name is given to the type of point mutation shown in Figure 4.11?

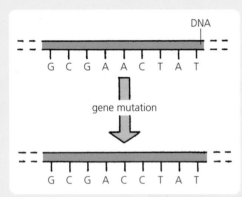

Figure 4.11

10 What name is given to the type of point mutation illustrated in Figure 4.12?

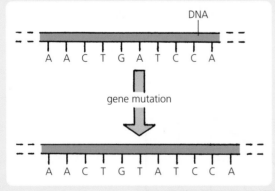

Figure 4.12

11 Which of the following are point mutations that BOTH lead to a major change that causes a large portion of the gene's DNA to be misread?

    A deletion and insertion   **B** insertion and translocation

    C translocation and substitution   **D** substitution and deletion

12 If a mutation occurs at a splice site on a chromosome, the codon for an intron–exon splice may be affected and, in error, an

    A intron may be retained by the mature mRNA transcript.

    B exon may be retained by the mature mRNA transcript.

    C intron may be retained by the primary mRNA transcript.

    D exon may be retained by the primary mRNA transcript.

13 Neurofibromatosis, a condition in which the human sufferer develops multiple brown lumps in the skin, is caused by a dominant mutant allele whose mutation frequency is 100 per million gametes. The chance of a new mutation occurring is therefore

    A 1 in 1000   **B** 1 in 10 000   **C** 1 in 100 000   **D** 1 in 1000 000

Questions 14 and 15 refer to Figure 4.13. It shows the amino acid sequences that belong to a particular polypeptide chain present in the wild type variety and several mutant strains of a species of bacterium.

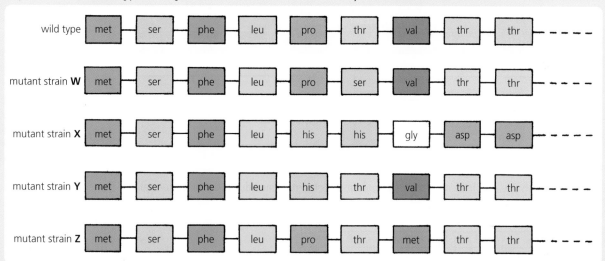

Figure 4.13

14 Which mutant strain has undergone an insertion of a nucleotide into its DNA sequence?

    A W   **B** X   **C** Y   **D** Z

15 The polypeptide made by mutant strain Y was found to function in a different way from that made by the wild type strain. This type of mutation is referred to as

    A trinucleotide repeat expansion.   **B** splice-site.

    C nonsense.   **D** missense.

16 Phenylketonuria is a genetic disorder. If its frequency of incidence is 0.1 per 1000 births, how many sufferers would be found among a population of 60 million people?

    A 60   **B** 600   **C** 6000   **D** 60 000

17 Figure 4.14 shows the outcome of a cross between sufferers of sickle cell trait (where H = allele for normal haemoglobin and S = allele for haemoglobin S).

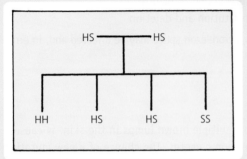

Figure 4.14

With respect to survival of the offspring, which row in Table 4.3 would, on average, be MOST likely for a large population?

	% number of survivors	
	population living in malarial area	population living in non-malarial area
A	25	75
B	25	100
C	50	75
D	50	100

Table 4.3

18 Figure 4.15 shows a pair of matching (homologous) chromosomes during gamete formation.

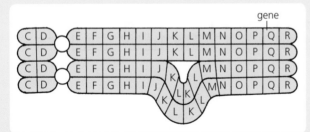

Figure 4.15

Which of the following terms refers to the type of mutation that has affected the altered chromosome?

A duplication   B substitution   C deletion   D translocation

Questions 19 and 20 refer to Figure 4.16, where chromosomes 1 and 2 are undergoing mutations.

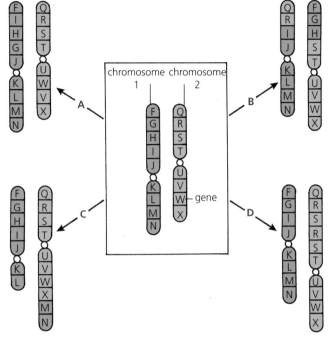

Figure 4.16

19 Which arrow represents reciprocal translocation?
20 Which arrow represents non-reciprocal translocation?

5 Human genomics

Matching Test
Match the terms in list X with their descriptions in list Y.

list X
1 amplification
2 bioinformatics
3 cooling
4 DNA polymerase
5 fluorescent labelling
6 genomics
7 microarray
8 neutral
9 pharmacogenetics
10 polymerase chain reaction (PCR)
11 primer
12 probe
13 sequencing
14 specific target sequence
15 systematics

list Y
a) process that allows primers to bind to target sequences during PCR
b) study of the effects of pharmaceutical drugs on the genetically diverse members of a population
c) laboratory technique used to create many copies of a piece of DNA
d) short sequence of nucleotides needed by DNA polymerase to begin replication of DNA
e) enzyme required to promote DNA replication
f) region at the end of a DNA strand complementary to a primer
g) application of statistics and computer technology to analyse and compare genetic sequence data
h) determining the order of bases on DNA fragments and the order of the fragments in a genome
i) increase in the number of copies of a DNA molecule by PCR
j) study of genomes
k) study of a group of living things with respect to their diversity, relatedness and classification
l) type of mutation that has no negative effect on the organism
m) orderly arrangement of thousands of different DNA probes attached to a glass slide
n) short, single-stranded fragment of DNA used to detect a specific sequence of nucleotide bases
o) technique used to indicate those spots on an array where a probe has combined with its complementary sequence of DNA

Multiple Choice Test
Choose the ONE correct answer to each of the following multiple choice questions.

1 Figure 5.1 shows the DNA fragments that resulted from two copies of part of a gene, each cut by a different enzyme. The computer works out the sequence by looking for matching overlaps between fragments. It found five of them to possess matching genetic sequences as indicated by the coloured regions.

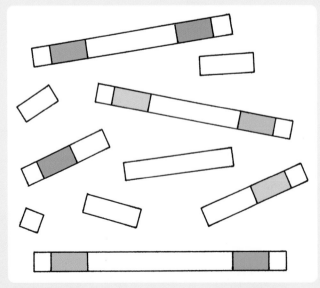

Figure 5.1

Which part of Figure 5.2 indicates the correct genomic sequence?

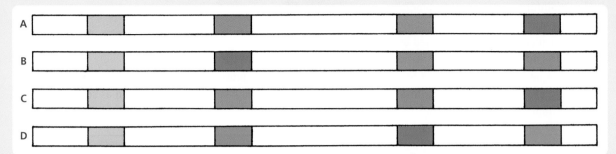

Figure 5.2

Questions 2 and 3 refer to the following information and to the possible answers that follow.

The DNA fragments shown in Figure 5.4 were formed during a type of sequencing where each fluorescent tag indicates the point on the strand where replication of complementary DNA was brought to a halt by a modified nucleotide.

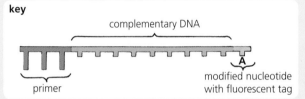

Figure 5.3

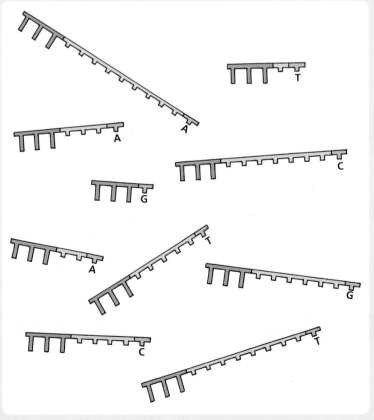

Figure 5.4

A GTAACTGCTA	B TCGTCAATGA	C CATTGACGAT	D AGCAGTTACT

2 Which is the sequence of bases in the complementary DNA strand?

3 Which is the sequence of bases in the original DNA strand?

4   Which of the following statements is NOT correct?

   During bioinformatics, computer programs are used to identify

   A   short sequences that contain a series of stop codons.

   B   start sequences that are followed by coding sequences.

   C   protein-coding sequences that are similar to those present in genes.

   D   base sequences that correspond to the amino acid sequence of a protein.

5   Which part of Figure 5.5 indicates, in a simple way, the pattern of very early human migration as supported by evidence from genomic sequencing?

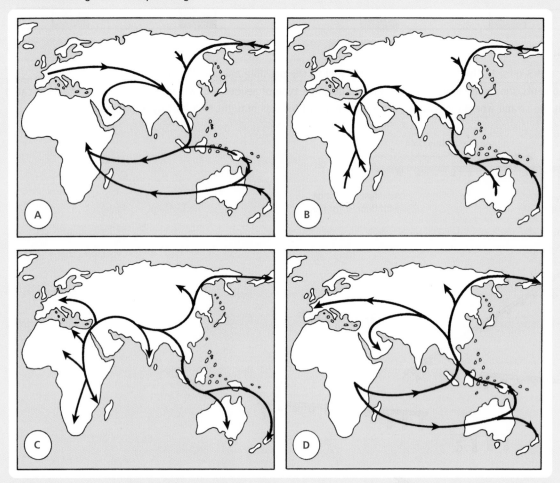

Figure 5.5

6   Table 5.1 gives the answers to the three blanks in the following sentence. Which row is correct?

   In the future, knowledge of the genetic components of diseases and the use of personal ___1___ may lead to ___2___ and each medical treatment will be customised to suit the requirements of an individual's ___3___ .

	Blank 1	Blank 2	Blank 3
A	metabolism	pharmacogenetics	genomics
B	genomics	pharmacogenetics	metabolism
C	metabolism	genomics	pharmacogenetics
D	genomics	metabolism	pharmacogenetics

Table 5.1

7 Figure 5.6 shows the events that lead to a genetic disorder. X, Y and Z indicate three means by which a rationally designed drug may be employed. Which of these would provide a therapeutic benefit?

   A X and Y only   **B** X and Z only

   C Y and Z only   **D** X, Y and Z

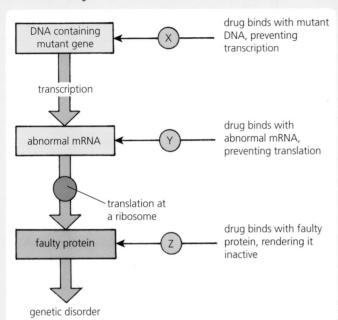

Figure 5.6

8 The steps given in the following list are involved in the first cycle of the PCR technique.

   1 Heat-tolerant DNA polymerase adds nucleotides to the primers.

   2 DNA to be amplified is heated to separate its two strands.

   3 Two identical copies of the original DNA molecule are formed.

   4 Following cooling, each primer binds to its target DNA sequence.

Their correct sequence is

   A 1,4,3,2   **B** 2,4,1,3   **C** 2,1,4,3   **D** 4,2,3,1

9 Figure 5.7 represents the early stages of the first cycle of PCR.

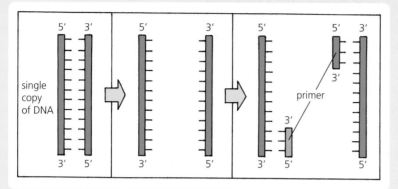

Figure 5.7

Which of the answers in Figure 5.8 shows the correct outcome at the end of the first cycle?

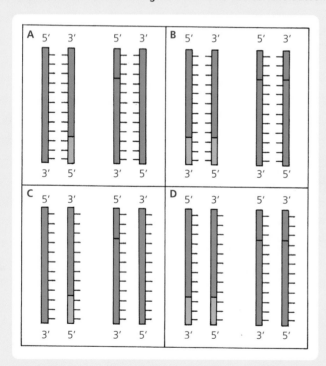

Figure 5.8

10 Starting with a single DNA molecule, how many DNA molecules would be present after eight cycles of the PCR procedure?

A 16   B 64   C 256   D 512

11 Figure 5.9 shows the expected number of copies of DNA that would be generated by the polymerase chain reaction (PCR) under ideal conditions.

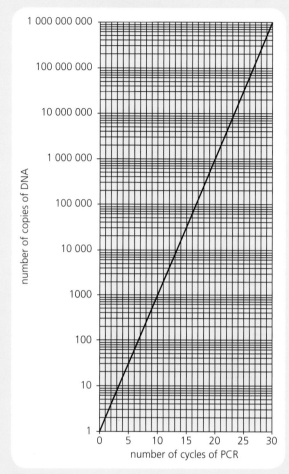

Figure 5.9

How many more cycles are required to increase the number of copies of DNA already present at five cycles by a factor of 10^6?

A 10   B 17   C 20   D 25 ➡

12 Figure 5.10 shows a DNA sample and four DNA probes (1, 2, 3 and 4) in a microarray.

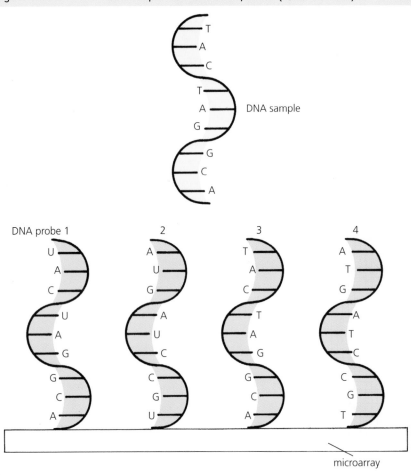

Figure 5.10

Which of the probes will detect the presence of the DNA sample?

A 1   B 2   C 3   D 4

13 On rare occasions a genetic test using a DNA probe fails to detect a specified DNA sequence when in fact the sequence was present. Such a result is called a

A false positive.   B false negative.   C positive control.   D negative control.

Questions 14 and 15 refer to Figure 5.11, which shows genetic fingerprints.

14 Identify person X's father.

15 Identify person X's sister.

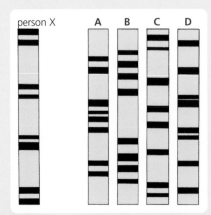

Figure 5.11

6 Metabolic pathways

Matching Test Part 1
Match the terms in list X with their descriptions in list Y.

list X

1 activation energy
2 active site
3 affinity
4 alternative route
5 anabolism
6 catabolism
7 catalyst
8 end product
9 enzyme
10 induced fit
11 metabolism
12 orientation
13 reaction rate
14 reversible
15 specificity
16 substrate
17 transition

list Y

a) substance that increases the rate of a chemical reaction and remains unaltered by the reaction
b) complementary relationship of molecular structure allowing an enzyme to combine with one type of substrate only
c) substance formed as a result of an enzyme acting on its substrate
d) sum of all the enzyme-controlled biochemical reactions occurring within a cell
e) term for a step in a metabolic pathway that can operate in both a forward and backward direction
f) step(s) in a metabolic pathway that allow the regular steps to be bypassed
g) substance upon which an enzyme acts, resulting in the formation of an end product
h) power needed to break the chemical bonds in the reactants in a chemical reaction
i) region of an enzyme molecule where the complementary surface of its substrate molecule becomes attached
j) type of metabolic pathway that brings about the breakdown of complex molecules and releases energy
k) degree of chemical attraction between reactant molecules
l) state of reactant molecules that have absorbed enough energy to break their bonds and allow the reaction to occur
m) protein made by living cells that acts as a biological catalyst
n) type of metabolic pathway that brings about the biosynthesis of complex molecules and requires energy
o) state of close molecular contact resulting from change in shape of an enzyme's active site to accommodate its substrate
p) way in which molecules of two reactants are held together, as determined by the shape of the enzyme's active site
q) amount of chemical change that occurs per unit time

Matching Test Part 2
Match the terms in list X with their descriptions in list Y.

list X

1 enzyme induction
2 inducer
3 operon
4 regulator gene
5 repressor
6 structural gene
7 transformation

list Y

a) alteration of a cell's genotype using a genetically modified plasmid
b) combination of an operator gene and one or more structural genes
c) region of DNA that codes for a repressor molecule
d) molecule that prevents a repressor molecule from combining with an operator gene
e) region of DNA that codes for a functional protein such as an enzyme
f) process by which an operator gene becomes free and an operon is able to code for an enzyme
g) molecule coded for by a regulator gene that can combine with an operator gene

Matching Test Part 3
Match the terms in list X with their descriptions in list Y.

list X
1 activator
2 active site
3 allosteric site
4 competitive
5 end-product inhibition
6 inhibitor
7 non-competitive
8 regulatory molecule
9 signal molecule

list Y
a) process by which a metabolite at a later stage in a pathway builds up and prevents the activity of an enzyme controlling an earlier stage
b) substance that acts on an enzyme as an activator or an inhibitor
c) regulatory molecule that becomes attached to an enzyme molecule, holding it in its active form
d) type of inhibitor that becomes attached to an enzyme at a position that is not the active site and changes the enzyme's molecular shape
e) region on an enzyme molecule to which the complementary surface of the substrate molecule becomes attached
f) type of inhibitor with a molecular structure similar to an enzyme's substrate, enabling it to become attached to the enzyme's active site
g) non-active location on an enzyme molecule to which an inhibitor or an activator may become attached
h) chemical substance (often from outside a cell) that exerts control over a metabolic pathway within the cell
i) regulatory molecule that decreases or halts the rate of an enzyme-controlled reaction

Multiple Choice Test
Choose the ONE correct answer to each of the following multiple choice questions.

1 Which of the following terms could NOT be used to describe correctly the type of pathway represented by the blue arrows in Figure 6.1?

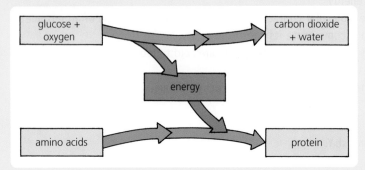

Figure 6.1

A metabolic   B anabolic   C biochemical   D catabolic

2 Which of the graphs in Figure 6.2 represents a chemical reaction controlled by a catalyst?

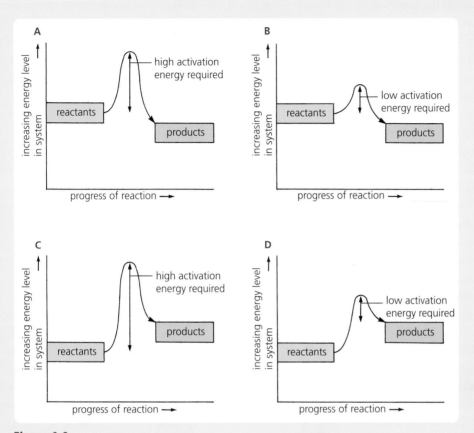

Figure 6.2

Questions 3 and 4 refer to Figure 6.3, which shows four stages that occur during an enzyme-controlled reaction.

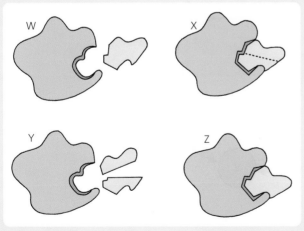

Figure 6.3

3 Which of the following indicates the correct sequence in which the four stages would occur if the enzyme promotes the building-up of a complex molecule from simpler ones?

A Y ⟶ X ⟶ Z ⟶ W   B W ⟶ Z ⟶ X ⟶ Y
C Y ⟶ Z ⟶ X ⟶ W   D W ⟶ X ⟶ Z ⟶ Y

4 Which stage(s) illustrate a state of induced fit between the enzyme and its substrate?

A X only   B Z only   C X and Z   D W, X and Z

5 Figure 6.4 charts the effect of increasing substrate concentration on the rate of an enzyme-controlled reaction where the concentration of enzyme is limited.

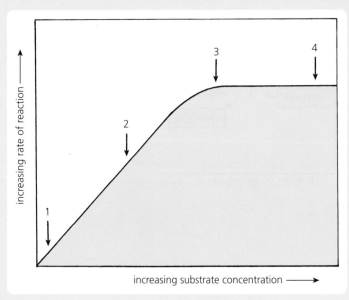

Figure 6.4

The concentration of substrate is the limiting factor at point(s)

A 1 only.   B 1 and 2.   C 3 only.   D 3 and 4.

6 Figure 6.5 shows a metabolic pathway where each encircled letter represents a metabolite.

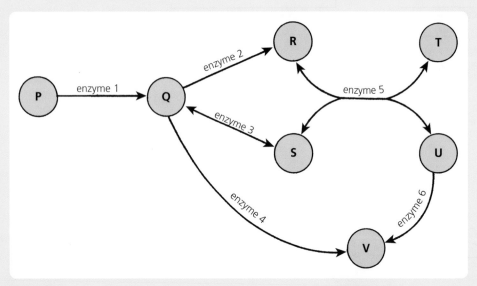

Figure 6.5

By which of the following alternative routes could a supply of metabolite V be obtained if enzyme 4 became inactive?

A   Q $\xrightarrow{\text{enzyme 2}}$ R $\xrightarrow{\text{enzyme 5}}$ U $\xrightarrow{\text{enzyme 6}}$ V

B   Q $\xrightarrow{\text{enzyme 3}}$ S $\xrightarrow{\text{enzyme 6}}$ U $\xrightarrow{\text{enzyme 5}}$ V

C   S $\xrightarrow{\text{enzyme 3}}$ Q $\xrightarrow{\text{enzyme 2}}$ R $\xrightarrow{\text{enzyme 6}}$ U

D   S $\xrightarrow{\text{enzyme 5}}$ T $\xrightarrow{\text{enzyme 6}}$ R $\xrightarrow{\text{enzyme 5}}$ U

7   Figure 6.6 shows part of a metabolic pathway under genetic control.

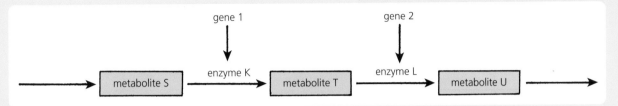

Figure 6.6

Which row in Table 6.1 correctly summarises the effect that a mutation could have on this pathway?

	Site of major genetic fault	Substance that accumulates as a result of an error of metabolism
A	1	S
B	1	T
C	2	L
D	2	U

Table 6.1

Questions 8 and 9 refer to Figure 6.7, which shows a possible arrangement of the genes involved in the induction of the enzyme β-galactosidase in the bacterium *Escherichia coli*.

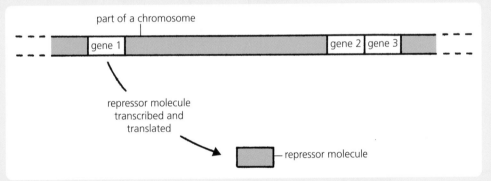

Figure 6.7

8   In the absence of lactose, the repressor molecule combines with gene 2 and, as a result, gene 3 remains 'switched off'. Which row in Table 6.2 indicates the correct identity of the three genes?

	Operator gene	Structural gene	Regulator gene
A	2	1	3
B	3	2	1
C	2	3	1
D	1	3	2

Table 6.2

9 Which of the following situations would arise if lactose became available to the cell?

	Gene 1	Gene 2	Gene 3
A	–	–	+
B	–	+	–
C	+	–	+
D	+	+	+

(+ = 'switched on', – = 'switched off')

Table 6.3

Questions 10 and 11 refer to the following information, diagram and table of possible answers.

Tryptophan is an amino acid needed for the synthesis of proteins. Situation 1 in Figure 6.8 shows a set of circumstances where a structural gene remains switched on in a cell of *Escherichia coli*. Under different circumstances, the series of events shown in situation 2 is thought to occur. This brings about the repression of the synthesis of an enzyme.

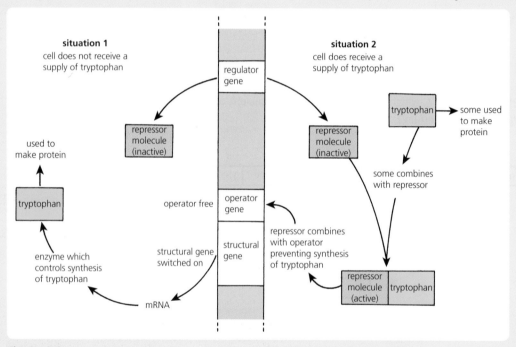

Figure 6.8

	Regulator gene	Operator gene	Structural gene
A	on	blocked	off
B	off	blocked	off
C	off	free	on
D	on	free	on

Table 6.4

10 Which answer in Table 6.4 refers to the state of the genes in a cell of *Escherichia coli* grown in nutrient broth lacking tryptophan?

11 Which answer refers to the state of the genes in a cell of *Escherichia coli* cultured in nutrient broth containing tryptophan?

12 Figure 6.9 shows an experiment set up to investigate the lac operon of *Escherichia coli*. ONPG is a colourless synthetic chemical that can be broken down by the enzyme β-galactosidase to form yellow compounds.

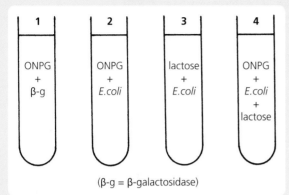

(β-g = β-galactosidase)

Figure 6.9

After an hour at 35°C a strong yellow colour will be formed in

A   tube 4 only.   B   tubes 1 and 4 only.

C   tubes 1, 2 and 4.   D   tubes 2, 3 and 4.

Questions 13, 14 and 15 refer to Figure 6.10, which represents the last four stages in a metabolic pathway in the fungus *Neurospora*.

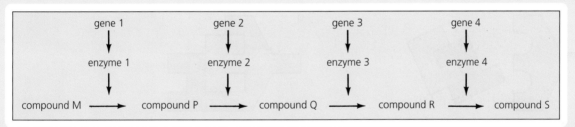

Figure 6.10

13 A mutant strain of the fungus is found to accumulate compound Q as a result of its metabolism. The gene that has undergone a mutation in this strain is

A 1   B 2   C 3   D 4

14 Wild type *Neurospora* can grow on minimal medium (sucrose, mineral salts and one vitamin) but mutant strains suffering a metabolic block are unable to do so. In an experiment the mutant strain referred to in question 13 was subcultured onto the following plates.

plate

P = minimal medium + substance P

Q = minimal medium + substance Q

R = minimal medium + substance R

S = minimal medium + substance S

This mutant strain would grow successfully on BOTH plates

A   P and Q.   B   Q and R.   C   R and S.   D   S and P.

15 A different mutant strain was found to grow successfully on plate S (minimal medium + substance S) but on no other. The enzyme that this strain of *Neurospora* fails to make is

A 1   B 2   C 3   D 4

16 Figure 6.11 shows the action of an enzyme on its substrate.

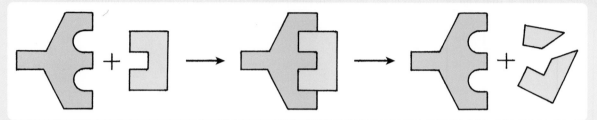

Figure 6.11

Which of the diagrams below shows how a competitive inhibitor brings about its effect on this system?

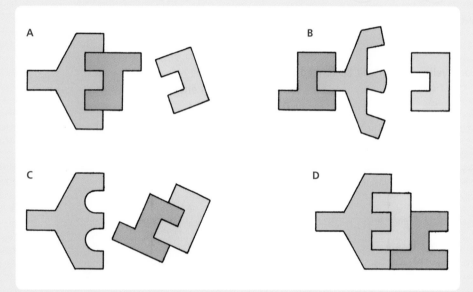

Figure 6.12

Questions 17 and 18 refer to Figure 6.13, which shows the effect of increasing substrate concentration on the rate of an enzyme-catalysed reaction affected by a limited amount of competitive inhibitor (and using a limited amount of enzyme).

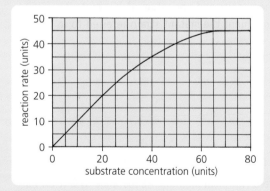

Figure 6.13

17 An increase in substrate concentration from 20 to 40 units brought about a percentage increase in reaction rate of

  A   15   B   35   C   55   D   75

18 At 70 units of substrate the active sites on the enzyme molecules would be occupied by

  A   mostly inhibitor molecules.   B   mostly substrate molecules.
  C   equal numbers of inhibitor and substrate molecules.   D   neither inhibitor nor substrate molecules.

19 Figure 6.14 shows an enzyme in its active and inactive forms.

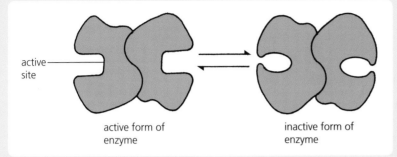

active site

active form of enzyme

inactive form of enzyme

Figure 6.14

Which of the diagrams in Figure 6.15 shows the action of a non-competitive inhibitor on this enzyme?

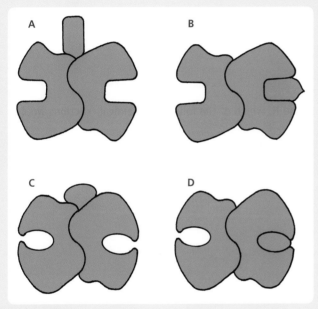

A

B

C

D

Figure 6.15

20 The red arrows in Figure 6.16, which shows a metabolic pathway, represent inhibition by an end product.

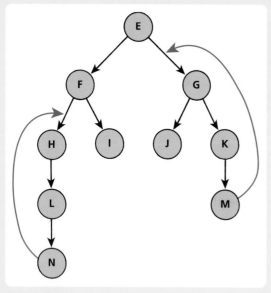

Figure 6.16

In the presence of high concentrations of metabolites M and N, which of the following chemical reactions would continue to proceed as before?

A E → G   B F → I   C G → J   D F → H

7 Cellular respiration & 8 Energy systems in muscle cells

Matching Test Part 1

Match the terms in list X with their descriptions in list Y.

list X	list Y
1 ADP	a) enzyme-controlled process by which a phosphate group is added to a molecule, often making it more reactive
2 anabolism	
3 ATP	b) role played by ATP between energy-releasing and energy-consuming reactions
4 ATP synthase	c) enzyme that removes hydrogen ions and high-energy electrons from the respiratory substrate
5 catabolism	
6 cellular respiration	d) respiratory substrate that provides energy for the regeneration of ATP
7 dehydrogenase	e) energy-consuming metabolic process, often in the form of a biosynthetic pathway
8 energy transfer	f) general name for a metabolic process involving the breakdown of complex to simpler molecules, normally releasing energy
9 glucose	
10 phosphorylation	g) low-energy molecule composed of adenosine and two phosphate groups
11 P_i	h) inorganic phosphate group needed to make ATP from ADP
	i) high-energy molecule composed of adenosine and three phosphate groups
	j) series of metabolic pathways that release energy from food, allowing ATP to be regenerated
	k) enzyme that catalyses the synthesis of ATP

Matching Test Part 2

Match the terms in list X with their descriptions in list Y.

list X	list Y
1 acetyl coenzyme A	a) the final hydrogen acceptor that combines with hydrogen ions (and electrons) to form water
2 citrate	
3 citric acid cycle	b) product of fermentation in animal cells
4 creatine phosphate	c) compound formed when glucose undergoes glycolysis
5 electron transport chain	d) enzyme that catalyses an irreversible step in glycolysis that acts as a key regulatory point
6 energy investment	
7 energy payoff	e) metabolite formed from oxaloacetate and an acetyl group
8 fast-twitch	f) coenzyme molecules that act as hydrogen acceptors
9 fermentation	g) location of electron transport chains in mitochondria
10 glycolysis	h) metabolic pathway that breaks glucose down into pyruvate
11 inner membrane	i) first phase of glycolysis, which requires ATP
12 lactate	j) compound present in muscle fibres that breaks down during strenuous activity, releasing energy used to make ATP
13 NAD and FAD	
14 oxaloacetate	k) compound formed from pyruvate and coenzyme A in the presence of oxygen
15 oxygen	l) circular metabolic pathway of stages controlled by enzymes that remove hydrogen ions from respiratory substrate
16 phosphofructokinase	
17 pyruvate	m) form of cellular respiration involving partial breakdown of glucose in the absence of oxygen
18 slow-twitch	
	n) type of muscle fibre that contracts and fatigues more slowly than fast-twitch fibres
	o) second phase of glycolysis, which generates ATP
	p) type of muscle fibre that contracts and fatigues more rapidly than slow-twitch fibres
	q) metabolite in citric acid cycle that combines with an acetyl group to form citrate
	r) group of protein molecules in a mitochondrial membrane that make energy available to pump hydrogen ions across the membrane

Multiple Choice Test

Choose the ONE correct answer to each of the following multiple choice questions.

1 Which of the following diagrams BEST represents the structure of a molecule of ATP (adenosine triphosphate)?

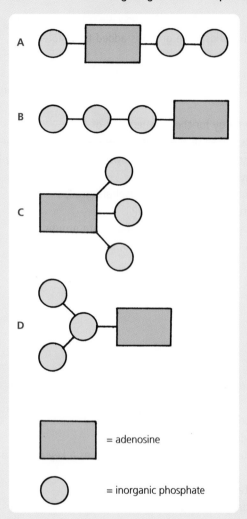

Figure 7.1

2 Which of the following equations represents the regeneration of ATP from its components?

A $\text{ADP} + \text{P}_i \xrightarrow{\text{energy taken in}} \text{ATP}$

B $\text{ADP} + \text{P}_i + \text{P}_i \xrightarrow{\text{energy taken in}} \text{ATP}$

C $\text{ADP} + \text{P}_i \xrightarrow{\text{energy released}} \text{ATP}$

D $\text{ADP} + \text{P}_i + \text{P}_i \xrightarrow{\text{energy released}} \text{ATP}$

3 Figure 7.2 represents a summary of cellular respiration.

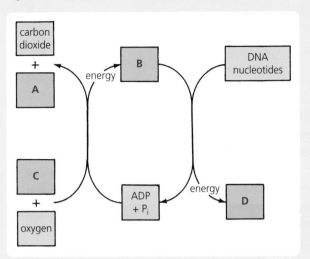

Figure 7.2

Which box represents the correct position of ATP in this scheme?

4 Which of the following is an example of anabolism?

A deamination of amino acids to form urea
B digestion of starch by enzymes
C formation of protein from amino acids
D oxidation of glucose during respiration

5 One mole of glucose releases 2880 kJ of energy. During aerobic respiration 44% of this is used to generate ATP. Therefore, the number of kilojoules per mole of glucose transferred to ATP is

A 126.7
B 161.3
C 1267.2
D 1612.8

6 The phosphorylation of glucose during glycolysis results in the production of

A low energy glucose-6-phosphate and ATP.
B low energy glucose-6-phosphate and ADP.
C high energy glucose-6-phosphate and ATP.
D high energy glucose-6-phosphate and ADP.

Questions 7 and 8 refer to the experiment shown in Figure 7.3. It was set up to investigate whether glucose-1-phosphate (G-1-P), a phosphorylated form of glucose, is more reactive than glucose, the non-phosphorylated form. Starch-free potato extract contains the enzyme phosphorylase, which promotes the synthesis of starch. The results were obtained by adding iodine solution to the cavities at 3-minute intervals.

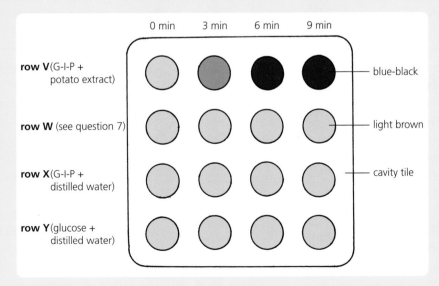

Figure 7.3

7 At the start of the experiment the cavities in row W should have received
  A glucose + potato extract.
  B glucose-1-phosphate + glucose.
  C glucose + iodine solution.
  D potato extract + distilled water.

8 If row X had not been set up as a negative control, then it would be valid to suggest that
  A G-1-P would have become phosphorylated in the presence of distilled water alone.
  B starch would have been formed whether or not the glucose was phosphorylated.
  C G-1-P would have become starch whether or not phosphorylase was present.
  D glucose would have become phosphorylated in the presence of phosphorylase.

9 Which row in Table 7.1 correctly refers to the effect that glycolysis of one molecule of glucose has on the number of molecules of ATP involved?

	Phase of glycolysis			
	Energy investment		Energy payoff	
	2 ATP used up	2 ATP generated	4 ATP used up	4 ATP generated
A		✓		✓
B		✓	✓	
C	✓		✓	
D	✓			✓

Table 7.1

Questions 10 and 11 refer to Figure 7.4, which shows the metabolic pathways leading to and including the citric acid cycle.

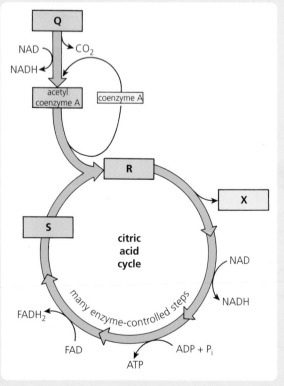

Figure 7.4

10 Which row in Table 7.2 indicates the correct identity of metabolites Q, R and S?

	Q	R	S
A	citrate	pyruvate	oxaloacetate
B	pyruvate	citrate	oxaloacetate
C	oxaloacetate	citrate	pyruvate
D	pyruvate	oxaloacetate	citrate

Table 7.2

11 The substance in box X should be

A NADH.   B ATP.   C CO_2.   D water.

12 The enzymes required for the citric acid cycle in a cell are located in the

A cytoplasmic fluid surrounding each mitochondrion.   B outer membrane of each mitochondrion.
C intermembrane space in each mitochondrion.   D central matrix of each mitochondrion.

Questions 13, 14 and 15 refer to Figure 7.5. It represents a section through the inner membrane of a mitochondrion and shows the processes involved in energy transfer.

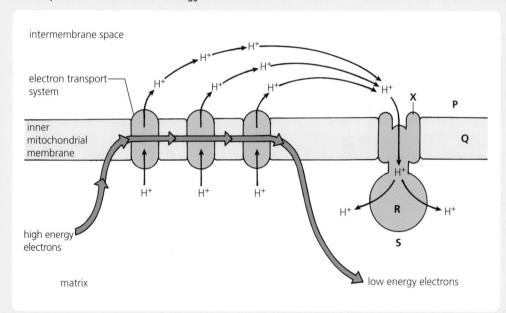

Figure 7.5

13 The region containing the highest concentration of hydrogen ions would be

A P.   B Q.   C R.   D S.

14 The function of molecule X is to

A promote energy release from ATP.   B act as an electron transport chain.
C pump hydrogen ions out of the matrix.   D catalyse the synthesis of ATP.

15 The final electron acceptor in this system is

A water.   B oxygen.   C ADP.   D NAD.

16 The enzymes that remove hydrogen ions from the respiratory substrate at several steps in glycolysis and the citric acid cycle are called

A decarboxylases.   B phosphorylases.
C dehydrogenases.   D synthases.

17 In an investigation into aerobic respiration in yeast cells, four test tubes were set up as indicated in Table 7.3 (where ✓ = present and X = absent).

Resazurin dye is a chemical that changes colour upon gaining hydrogen as follows:

blue ⟶ pink ⟶ colourless

(lacks hydrogen)　　　(some hydrogen gained)　　　(much hydrogen gained)

Substance	Test tube			
	1	**2**	**3**	**4**
live yeast	✓	X	✓	X
dead yeast	X	X	X	✓
glucose solution	X	✓	✓	✓
resazurin dye	✓	✓	✓	✓

Table 7.3

The disappearance of resazurin's blue colour will

A occur in tube 3 more quickly than in tube 1.
B occur in tube 2 more quickly than in tube 1.
C fail to occur in both tubes 1 and 2.
D fail to occur in both tubes 3 and 4.

18 During aerobic respiration of one molecule of glucose, MOST ATP is synthesised during

A the citric acid cycle.
B electron transport.
C glycolysis.
D the breakdown of oxaloacetate.

Questions 19 and 20 refer to Figure 7.6. It shows the relationship between carbohydrate and two other classes of food that can act as respiratory substrates.

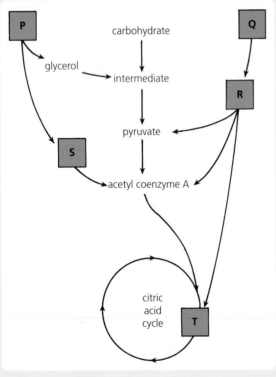

Figure 7.6

19 Which of the following boxes represents fats?

   A P   **B** Q   **C** S   **D** T

20 Which of the following boxes represents amino acids?

   A P   **B** R   **C** S   **D** T

21 Figure 7.7 shows the apparatus set up to investigate the use of sucrose as a respiratory substrate by yeast.

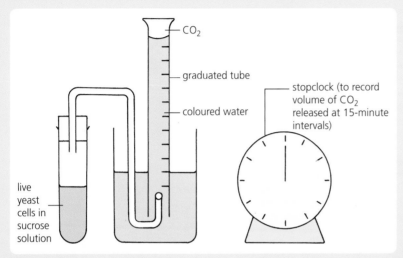

Figure 7.7

Which row in Table 7.4 is correct?

	Dependent variable	Independent variable
A	concentration of sucrose solution	time
B	volume of CO_2	concentration of sucrose solution
C	volume of CO_2	time
D	time	volume of CO_2

Table 7.4

22 Figure 7.8 shows the regulation of the respiratory pathway by feedback control at arrows X and Y.

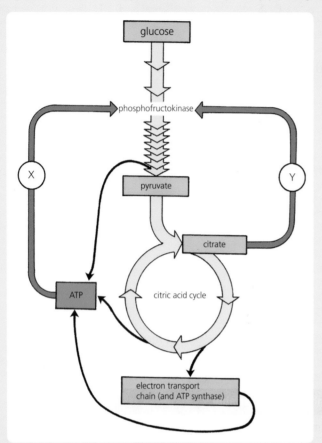

Figure 7.8

Which row in Table 7.5 is correct?

	Effect on respiratory pathway	
	Arrow X	**Arrow Y**
A	stimulatory	stimulatory
B	inhibitory	stimulatory
C	stimulatory	inhibitory
D	inhibitory	inhibitory

Table 7.5

23 Which of the following equations represents a chemical reaction that occurs for about 10 seconds in muscle cells during strenuous activity?

A   creatine phosphate + ADP ⟶ creatine + ATP

B   creatine phosphate + ATP ⟶ creatine + ADP

C   creatine + ADP ⟶ creatine phosphate + ATP

D   creatine + ATP ⟶ creatine phosphate + ADP

24 Figure 7.9 shows the process of fermentation in muscle cells lacking sufficient oxygen.

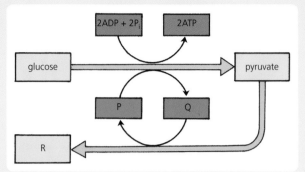

Figure 7.9

Which row in Table 7.6 correctly identifies the contents of boxes P, Q and R?

	Box P	Box Q	Box R
A	2 NAD	2 NADH	citrate
B	2 NADH	2 NAD	lactate
C	2 NADH	2 NAD	citrate
D	2 NAD	2 NADH	lactate

Table 7.6

25 Table 7.7 compares fast-twitch and slow-twitch muscle fibres. Which row is correct?

	Feature	Fast-twitch muscle fibre	Slow-twitch muscle fibre
A	relative number of mitochondria present	large	small
B	major storage fuels used	glycogen and creatine phosphate	fats
C	relative concentration of myoglobin present in cells	high	low
D	respiratory pathway(s) normally used to generate ATP	glycolysis and aerobic pathways	glycolysis only

Table 7.7

Physiology and Health

9 The structure and function of reproductive organs and gametes and their role in fertilisation & 10 Hormonal control of reproduction

Matching Test

Match the terms in list X with their descriptions in list Y.

list X
1 corpus luteum
2 follicle
3 follicle-stimulating hormone (FSH)
4 interstitial cells
5 interstitial cell-stimulating hormone (ICSH)
6 luteinising hormone (LH)
7 oestrogen
8 ovulation
9 progesterone
10 prostate gland
11 semen
12 seminal vesicle
13 seminiferous tubules
14 testosterone

list Y
a) site of sperm production in the testes
b) release of an ovum
c) hormone made in the testes that promotes sperm production
d) glandular structure that develops from a follicle after ovulation
e) pituitary hormone that brings about ovulation
f) structure that produces a component of semen containing enzymes that control its viscosity
g) site of testosterone production in the testes
h) milky liquid containing sperm and secretions from seminal vesicles and prostate gland
i) pituitary hormone that promotes sperm production in males and follicle maturation in females
j) ovarian hormone responsible for proliferation of the endometrium before ovulation
k) structure that surrounds a developing egg
l) pituitary hormone that stimulates interstitial cells to produce testosterone
m) ovarian hormone responsible for vascularisation of the endometrium after ovulation
n) structure that produces a component of semen that is rich in fructose

Multiple Choice Test

Choose the ONE correct answer to each of the following multiple choice questions.

Questions 1 and 2 refer to Figure 9.1, which shows the male reproductive system.

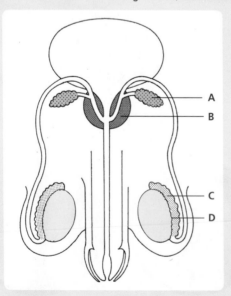

Figure 9.1

1 Which structure secretes a liquid rich in fructose?
2 Which structure secretes a lubricating liquid containing enzymes?

Questions 3 and 4 refer to Figure 9.2.

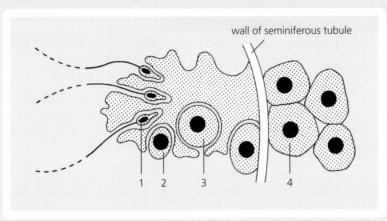

Figure 9.2

3 Figure 9.2 is a close-up of which region in Figure 9.1?
4 Testosterone is produced by cell type

   A 1   B 2   C 3   D 4

Questions 5 and 6 refer to the following possible answers.

   A to stimulate the testes to produce spermatozoa
   B to maintain optimum viscosity of liquid for sperm motility
   C to provide sperm with the energy needed for motility
   D to help sperm reach the oviducts by stimulating uterine contractions

5 Which answer explains the presence in semen of fructose?
6 Which answer explains the presence in semen of hormone-like substances produced by the seminal vesicles?
7 Figure 9.3 shows the self-regulating effect of testosterone.

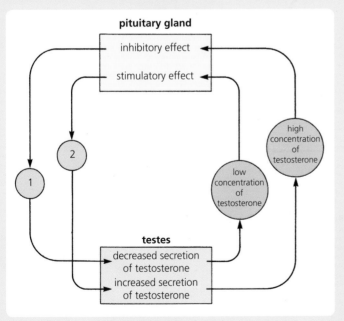

Figure 9.3

Which row in Table 9.1 correctly identifies the terms missing from circles 1 and 2?

	Circle 1	Circle 2
A	increased secretion of ICSH	decreased secretion of ICSH
B	decreased secretion of ICSH	increased secretion of ICSH
C	increased secretion of FSH	decreased secretion of FSH
D	decreased secretion of FSH	increased secretion of FSH

(ICSH = interstitial cell-stimulating hormone and FSH = follicle-stimulating hormone)

Table 9.1

Questions 8, 9 and 10 refer to Figure 9.4, which shows the relationship between the four hormones involved in the menstrual cycle and their target sites.

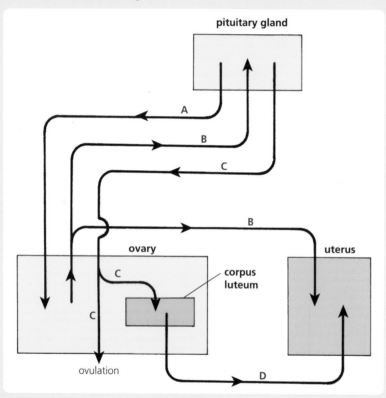

Figure 9.4

8 Which letter represents progesterone?

9 Which letter represents follicle-stimulating hormone (FSH)?

10 Which letter represents oestrogen?

11 The corpus luteum is a structure that

   A   secretes the hormone progesterone.

   B   releases an egg from the ovary.

   C   develops into a follicle.

   D   produces luteinising hormone.

12 The following list gives some of the effects brought about directly by hormones.
   1 promotion of sperm production
   2 stimulation of oestrogen production
   3 development of the corpus luteum
   4 maturation of a follicle
   5 stimulation of progesterone production
   Which of these are ALL affected by follicle-stimulating hormone (FSH)?
   A 1, 2 and 4   **B** 1, 3 and 5   **C** 2, 3 and 4   **D** 2, 4 and 5

Questions 13 and 14 refer to Figure 9.5, which shows relative concentrations of the four hormones present in the blood plasma of a woman during a menstrual cycle (where the first day of menstruation was regarded as the start of the cycle).

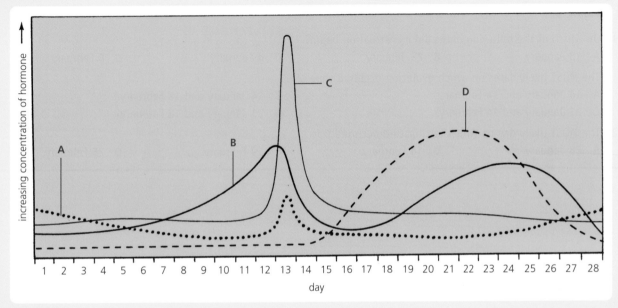

Figure 9.5

13 Which line on the graph represents luteinising hormone (LH)?
14 Which line on the graph represents oestrogen?
15 Which of the following are BOTH pituitary hormones?
   A follicle-stimulating hormone and luteinising hormone
   B luteinising hormone and progesterone
   C progesterone and oestrogen
   D oestrogen and follicle-stimulating hormone

16 During the last week of the luteal phase of the menstrual cycle the following events occur.
   1 A rapid drop in progesterone level takes place.
   2 Menstruation begins.
   3 Lack of LH leads to the degeneration of the corpus luteum.
   4 The endometrium is no longer maintained.
   The correct order in which these occur is
   A 1, 3, 4, 2   **B** 3, 1, 4, 2   **C** 1, 3, 2, 4   **D** 3, 1, 2, 4

17 Follicle-stimulating hormone (FSH) was administered to a group of rats that had had their anterior pituitary glands removed. Compared with the control group of normal rats, which of the following events did NOT take place in the experimental animals?
   A proliferation of the endometrium   **B** maturation of a follicle
   C development of the corpus luteum   **D** build-up of oestrogen in the bloodstream

Questions 18, 19 and 20 refer to Figure 9.6, which shows the changes that occurred in a woman's endometrium over a period of 63 days.

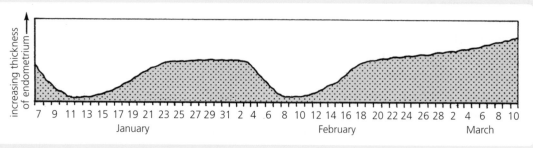

Figure 9.6

18 On which of the following dates did menstruation begin?

   A 13 January   **B** 23 January   **C** 4 February   **D** 8 February

19 The MOST likely dates on which ovulation occurred were

   A 14 January and 14 February.   **B** 14 January and 18 February.

   C 21 January and 14 February.   **D** 21 January and 18 February.

20 The MOST likely day on which fertilisation occurred was

   A 23 January.   **B** 14 February.   **C** 19 February.   **D** 25 February.

11 Biology of controlling fertility

Matching Test
Match the terms in list X with their descriptions in list Y.

list X
1 artificial insemination
2 barrier
3 chemical
4 continuous
5 cyclical
6 intra-cytoplasmic sperm injection (ICSI)
7 *in vitro* fertilisation (IVF)
8 pre-implantation genetic diagnosis (PGD)
9 sperm count
10 super-ovulation

list Y
a) term describing type of fertility where a brief fertile period alternates with a long infertile period every four weeks
b) term describing type of fertility where sperm are produced without interruption in response to the presence of testosterone
c) simultaneous release of a relatively large number of eggs following stimulation of ovaries by drugs that mimic FSH and LH
d) number of gametes present in a sample of semen
e) procedure used during IVF to identify single gene and chromosomal abnormalities in cells from embryos
f) method of contraception involving synthetic hormones that exert negative feedback, preventing the release of FSH and LH
g) method of contraception using a device that physically blocks sperm from reaching an ovum
h) introduction of semen into the female reproductive tract by means other than sexual intercourse
i) procedure involving fertilisation outside the bodies of the would-be parents
j) process by which the head of a sperm is drawn into a needle and then injected into an egg

Multiple Choice Test
Choose the ONE correct answer to each of the following multiple choice questions.

1 Which row in Table 11.1 is correct?

	Type of fertility in women	Hormone that brings about ovulation
A	continuous	luteinising hormone
B	cyclical	luteinising hormone
C	continuous	oestrogen
D	cyclical	oestrogen

Table 11.1

2 Which row in Table 11.2 is correct?

	Type of fertility in men	Hormone that stimulates sperm production directly
A	continuous	testosterone
B	cyclical	testosterone
C	continuous	interstitial cell-stimulating hormone
D	cyclical	interstitial cell-stimulating hormone

Table 11.2

Questions 3 and 4 refer to the chart in Figure 11.1, which records the changes in one woman's body temperature during a menstrual cycle.

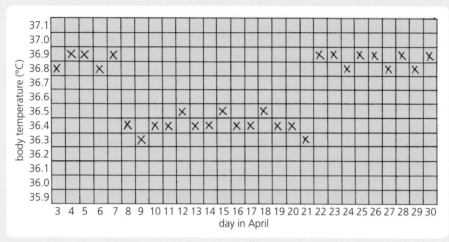

Figure 11.1

3 On which of the following days in April was the woman's cervical mucus thin and watery?

A 4   B 8   C 21   D 30

4 If sperm can survive for three days after intercourse and an egg can live for two days after ovulation, between which of the following days in April could intercourse have resulted in pregnancy?

A 6–10   B 12–16   C 19–23   D 26–30

5 A certain fertility clinic considers a man to be fertile if:

i) over 20 million sperm are present in 1 cm^3 of his semen;

ii) at least 40% of his sperm are active;

iii) at least 60% of his sperm are normal.

Which patient in Table 11.3 fails to meet these criteria fully?

	Patient			
	A	**B**	**C**	**D**
total number of sperm (millions/cm^3)	23	29	35	41
number of active sperm (millions/cm^3)	11	13	14	16
number of normal sperm (millions/cm^3)	16	18	21	24

Table 11.3

6 Figure 11.2 shows the reproductive systems of four women (P, Q, R and S).

P
partial blockage
Q
complete blockage
R
complete blockage
S
oviduct cut and tied

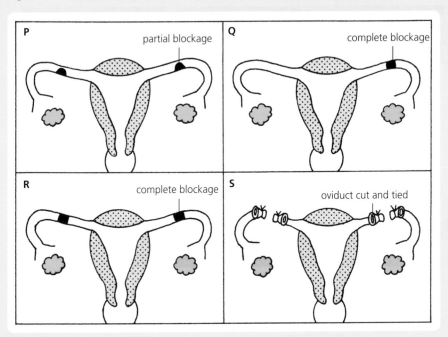

Figure 11.2

In which two women is fertilisation impossible at present?

A P and Q B Q and R C R and S D S and P

7 The introduction of semen into a woman's reproductive tract by some means other than sexual intercourse is called

A contraceptive implantation. B artificial insemination.

C *in vitro* fertilisation. D intracytoplasmic sperm injection.

8 Figure 11.3 gives some of the steps in the procedure employed during *in vitro* fertilisation (IVF).

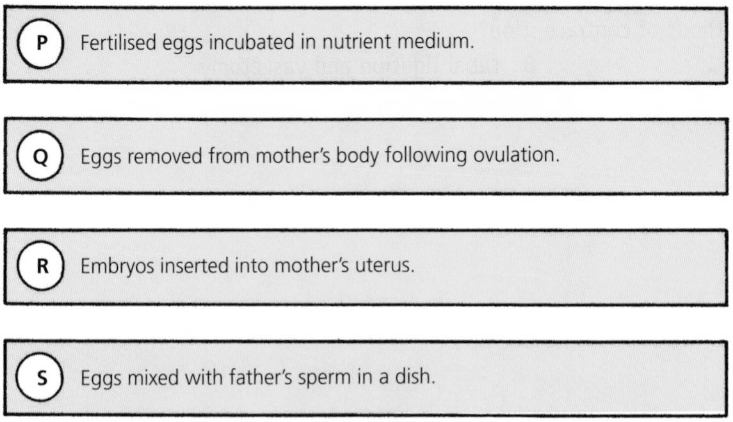

P Fertilised eggs incubated in nutrient medium.
Q Eggs removed from mother's body following ovulation.
R Embryos inserted into mother's uterus.
S Eggs mixed with father's sperm in a dish.

Figure 11.3

The order in which these steps are carried out is

A Q, S, P, R. B S, Q, P, R. C Q, S, R, P. D S, Q, R, P.

9 Which row in Table 11.4 correctly identifies two features of pre-implantation genetic diagnosis (PGD)?

	Specific approach	Non-specific approach	Checks for single gene disorders and common chromosomal abnormalities	Checks for a known chromosomal or gene defect
A	✓			✓
B	✓		✓	
C		✓	✓	
D		✓		✓

Table 11.4

10 Table 11.5 shows data that refer to *in vitro* fertilisation (IVF) success rates for a northern European country.

Age of patient (years)	IVF success rate as live birth rate per cycle (%)		
	2010	2011	2012
under 35	32.7	33.2	33.6
35–37	27.9	28.1	28.2
38–40	19.1	19.2	19.3
41–43	8.6	8.8	8.9
44 and over	3.3	3.4	3.5

Table 11.5

The chance of successfully giving birth following IVF treatment is
A increasing with time but decreases with increasing age of patient.
B increasing with time and increases with increasing age of patient.
C decreasing with time but increases with decreasing age of patient.
D decreasing with time and decreases with decreasing age of patient.

11 Which of the following are BOTH barrier methods of contraception?
A diaphragm and tubal ligation
B tubal ligation and vasectomy
C vasectomy and cervical cap
D cervical cap and diaphragm

12 Figure 11.4 shows an intra-uterine device (IUD).

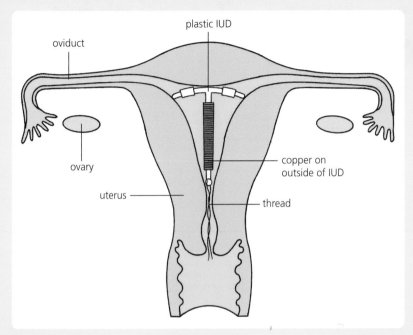

plastic IUD

oviduct

ovary

copper on
outside of IUD

uterus

thread

Figure 11.4

Which of the following is NOT one of the ways in which this physical appliance works to prevent pregnancy?

A It impairs the mobility of sperm, preventing them from reaching the egg.

B It prevents ovulation by negative feedback control involving hormones.

C Its presence stimulates the release of white blood cells that are hostile to sperm.

D It irritates the uterus lining making it unreceptive to the implantation of an embryo.

13 Oral contraceptive pills often contain synthetic versions of two hormones. Which column in Table 11.6 correctly identifies BOTH of them?

	A	B	C	D
oestrogen	✓	✓		
follicle-stimulating hormone	✓		✓	
progesterone		✓		✓
luteinising hormone			✓	✓

Table 11.6

14 The hormones in contraceptive pills work by exerting an inhibitory effect directly on the

A ovaries.　　　　B endometrium.　　　　C follicle.　　　　D pituitary gland.　➡

15 Which row in Table 11.7 gives the correct answers to blanks 1, 2 and 3 in the following sentence?

The hormone in a contraceptive implant is a synthetic version of _____1_____ and it works by exerting _____2_____ feedback control and inhibiting _____3_____.

	Blank 1	Blank 2	Blank 3
A	progesterone	positive	menstruation
B	luteinising hormone	positive	ovulation
C	progesterone	negative	ovulation
D	luteinising hormone	negative	menstruation

Table 11.7

12 Ante- and postnatal screening

Matching Test
Match the terms in list X with their descriptions in list Y.

list X
1. amniocentesis
2. anomaly scan
3. antenatal
4. anti-Rhesus antibodies
5. autosomal dominant
6. autosomal recessive
7. chorionic villus sampling (CVS)
8. dating scan
9. diagnosis
10. false negative
11. false positive
12. karyotyping
13. pedigree analysis
14. phenylketonuria (PKU)
15. postnatal
16. Rhesus antigens
17. sensitising event
18. sex-linked recessive

list Y
a) term referring to factors occurring or present before birth
b) term referring to factors occurring or present after birth
c) examining a photomicrograph of a complement of chromosomes to establish their size and number in pairs
d) form of ultrasound imaging carried out at 18–20 weeks of pregnancy to check for physical abnormalities in the fetus
e) form of ultrasound imaging carried out at 8–14 weeks of pregnancy to calculate when the baby is due to be born
f) result of a test that shows the fetus to have a certain condition when in fact it does not have it
g) result of a test that shows the fetus does not have a certain condition when in fact it does have it
h) inborn error of metabolism where the untreated sufferer accumulates excess phenylalanine
i) removal of amniotic fluid containing fetal cells for examination of chromosomes
j) removal of placental cells for examination of chromosomes
k) molecules on the surfaces of the red blood cells of a Rhesus-positive fetus
l) molecules given to a Rhesus-negative mother following a sensitising event at birth of a Rhesus-positive baby
m) situation where a Rhesus-negative mother's immune system comes in contact with Rhesus antigens from her Rhesus-positive baby
n) type of definitive test giving results that establish without doubt the presence or absence of a certain condition
o) study of a family tree with respect to an inherited trait
p) pattern of inheritance where males and females are affected equally and every affected person has an affected parent
q) pattern of inheritance where females are rarely or never directly affected
r) pattern of inheritance where the characteristic tends to miss a generation but affects both sexes equally

Multiple Choice Test
Choose the ONE correct answer to each of the following multiple choice questions.

1. Table 12.1 refers to features of scans carried out on pregnant women using ultrasound imaging. Which row in the table describes an anomaly scan?

	Carried out at 8–14 weeks of gestation	Carried out at 18–20 weeks of gestation	Used to check for abnormalities in the fetus	Used to calculate when the baby will be born
A	✓			✓
B		✓	✓	
C		✓		✓
D	✓		✓	

Table 12.1

2 Pre-eclampsia is a serious medical condition that affects a minority of pregnant women. Figure 12.1 shows the results of a survey carried out on a large population of women. (The error bars indicate a 95% level of confidence.)

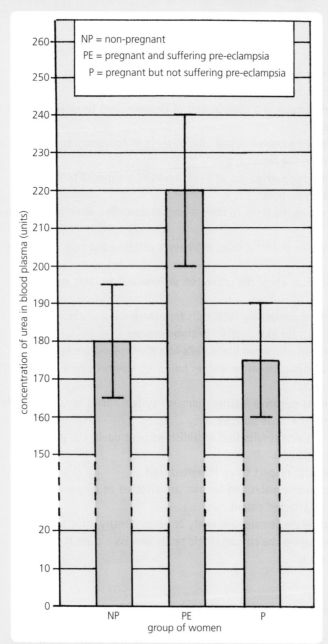

NP = non-pregnant
PE = pregnant and suffering pre-eclampsia
P = pregnant but not suffering pre-eclampsia

Figure 12.1

With a 95% level of confidence it can be concluded that the

A concentration of urea for NP exceeds that of P by 20 units.

B concentration of urea for PE exceeds that of P by 50 units.

C difference in concentration of urea between NP and P is statistically significant.

D difference in concentration of urea between NP and PE is statistically significant.

3 Figure 12.2 shows the mean level of human chorionic gonadotrophin (HCG) present in the blood of pregnant women. HCG is a marker chemical that may indicate the presence of Down's syndrome in a fetus.

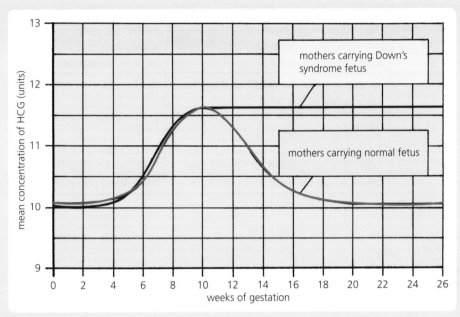

Figure 12.2

Risk management would be meaningless if it were based on a result of an HCG test carried out at

A 10 weeks.   B 15 weeks.   C 20 weeks.   D 25 weeks.

4 Table 12.2 shows data from an investigation into the effectiveness of ultrasound imaging on the antenatal detection of inherited malformations in babies born in a region of the UK.

Row	Category	Year		
		2006	2010	2014
1	total number of births	32 947	29 583	27 312
2	number of babies born with malformations	778	698	652
3	number of babies reported antenatally as having malformations but normal at birth	107	56	45
4	number of babies reported antenatally as being normal but born with malformations	458	392	347

Table 12.2

Which row in Table 12.3 is correct about the results in rows 3 and 4 of Table 12.2?

	False positive	False negative
A	row 3	row 4
B	row 4	row 3
C	rows 3 and 4	neither row
D	neither row	rows 3 and 4

Table 12.3

5 Table 12.4 compares two procedures carried out during prenatal diagnosis of a range of conditions. Which row in the table is NOT correct?

	Question	Procedure	
		chorionic villus sampling	**amniocentesis**
A	When is it carried out?	earlier in pregnancy	later in pregnancy
B	What is extracted?	sample of placental cells that have the same genes as the fetus	fluid containing fetal cells
C	What is the relative risk of it causing a miscarriage?	higher	lower
D	When can the cells be inspected?	cells must be cultured for several days before inspection	cells can be inspected immediately on extraction

Table 12.4

6 Which row in Table 12.5 indicates a situation where the mother will need to receive anti-Rhesus antibodies following a sensitising event at the birth of her baby?

	Genetics of mother	**Genetics of her newborn baby**
A	Rhesus negative	Rhesus negative
B	Rhesus positive	Rhesus negative
C	Rhesus negative	Rhesus positive
D	Rhesus positive	Rhesus positive

Table 12.5

7 In the UK all newborn babies are routinely screened for phenylketonuria (PKU), a genetic disorder that results in an error of metabolism. PKU sufferers are then
  A placed on a diet containing minimum phenylalanine.
  B placed on a diet containing maximum phenylalanine.
  C given medication containing the enzyme that they are unable to produce.
  D given medication to inactivate a harmful enzyme that they produce.

Questions 8 and 9 refer to the following possible answers.
  A A sufferer of the trait always has an affected parent.
  B The trait is never passed from a father to his sons.
  C Males and females are affected equally by the trait.
  D The trait tends to skip generations in a family pedigree.

8 Which characteristic applies SOLELY to a pattern of sex-linked recessive inheritance?
9 Which characteristic applies SOLELY to a pattern of autosomal dominant inheritance?

Questions 10 and 11 refer to Figure 12.3, which shows a family tree.

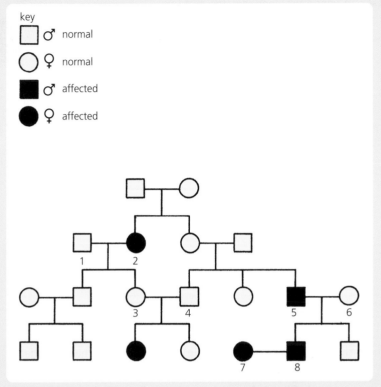

Figure 12.3

10 The pattern of inheritance shown by this trait is described as
   A sex-linked recessive.   B sex-linked dominant.   C autosomal recessive.   D autosomal dominant.

11 A consanguineous marriage took place between persons
   A 1 and 2.   B 3 and 4.   C 5 and 6.   D 7 and 8.

12 Achondroplasia is a form of restricted growth caused by a dominant allele. Figure 12.4 shows two family trees affected by this trait.

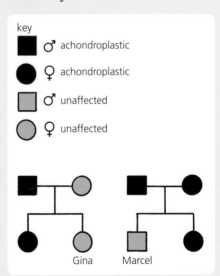

Figure 12.4

Gina and Marcel intend to marry. Provided that a spontaneous mutation does not occur, the chance of each of their children being affected by achondroplasia is
   A 0 in 4.   B 1 in 4.   C 2 in 4.   D 3 in 4.

Questions 13 and 14 refer to Figure 12.5, which shows a family pedigree for haemophilia. This trait shows a sex-linked recessive pattern of inheritance and sufferers lack an important blood-clotting protein.

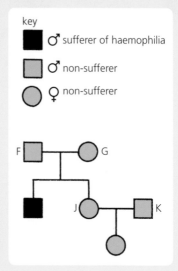

key
- ♂ sufferer of haemophilia
- ♂ non-sufferer
- ♀ non-sufferer

Figure 12.5

13 The chance of any daughter born to parents F and G being heterozygous for the trait is

   A 1 in 1.   **B** 1 in 2.   **C** 1 in 3.   **D** 1 in 4.

14 If couple J and K have a son, the chance (from the information so far available) that he will be a haemophiliac is

   A 1 in 1.   **B** 1 in 2.   **C** 1 in 3.   **D** 1 in 4.

Question 15 and 16 refer to the following information. Cystic fibrosis is an inherited disorder involving the secretion of abnormally thick mucus. It shows a recessive autosomal pattern of inheritance. In the family pedigree in Figure 12.6, N = dominant allele and n = recessive allele.

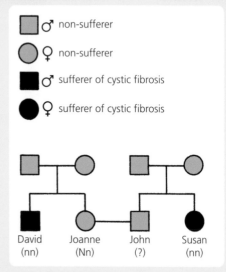

- ♂ non-sufferer
- ♀ non-sufferer
- ♂ sufferer of cystic fibrosis
- ♀ sufferer of cystic fibrosis

David (nn)   Joanne (Nn)   John (?)   Susan (nn)

Figure 12.6

15 From the information given in Figure 12.6, the chance of allele n being present in John's genotype is

   A 1 in 1.   **B** 1 in 2.   **C** 2 in 3.   **D** 3 in 4.

16 If John and Joanne have the same genotype, the chance of each of their children being a sufferer of cystic fibrosis is

   A 1 in 1.   **B** 1 in 2.   **C** 1 in 3.   **D** 1 in 4.

17 Figure 12.7 shows two family trees with a history of Huntington's chorea. This autosomal dominant trait leads to degeneration of the nervous system when the sufferer is in his/her late thirties.

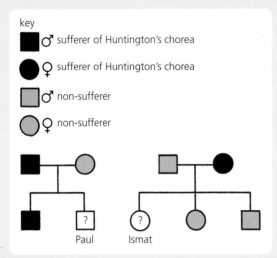

key

◼ ♂ sufferer of Huntington's chorea

● ♀ sufferer of Huntington's chorea

▨ ♂ non-sufferer

◯ ♀ non-sufferer

Figure 12.7

Paul and Ismat, both aged 23 years, intend to marry. They seek the advice of a genetic counsellor. She advises them that they could both be future sufferers of the disorder. If this turns out to be the case then the percentage, on average, of their children also being sufferers will be

A   25%   B   50%   C   75%   D   100%

Questions 18 and 19 refer to the following information. Hypophosphataemia is a disorder involving defective reabsorption of phosphate from the filtrate in the kidneys. It shows a sex-linked dominant pattern of inheritance as in the family pedigree in Figure 12.8.

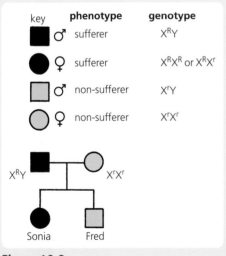

key	phenotype	genotype
◼ ♂	sufferer	X^RY
● ♀	sufferer	X^RX^R or X^RX^r
▨ ♂	non-sufferer	X^rY
◯ ♀	non-sufferer	X^rX^r

Figure 12.8

18 Which row in Table 12.6 gives the correct genotypes of Sonia and Fred?

	Sonia	Fred
A	X^RX^R	X^RY
B	X^RX^r	X^RY
C	X^RX^r	X^rY
D	X^RX^R	X^rY

Table 12.6

19 If Fred marries a woman with the same genotype as his sister, the chance of each of their daughters being affected by hypophosphataemia is

A 1 in 1   B 1 in 2   C 1 in 3   D 1 in 4

20 Table 12.7 refers to four disorders that show polygenic patterns of inheritance. Which of these disorders carries a 1 in 4 risk of an affected parent having a second affected child?

| | | Disorder | | |
| | A | B | C | D |
	epilepsy	mental disability	dislocation of hip	pyloric stenosis
incidence (per 100)	5	4	0.7	3
risk of normal parents having second affected child (per 100)	5	4	6	6
risk of affected parent having: an affected child (per 100)	5	10	12	10
risk of affected parent having: a second affected child (per 100)	10	20	33	25

Table 12.7

13 The structure and function of arteries, capillaries and veins

Matching Test

Match the terms in list X with their descriptions in list Y.

list X

1 arteriole
2 artery
3 capillary
4 elastic fibres
5 endothelium
6 lumen
7 lymph
8 lymph vessel
9 lymphatic duct
10 plasma
11 pressure filtration
12 smooth muscle
13 tissue fluid
14 valve
15 vasoconstriction
16 vasodilation
17 vein

list Y

a) thin layer of epithelial cells lining the lumen of a blood vessel
b) central cavity of a blood vessel
c) structures that enable the walls of an artery to stretch and recoil during surges of blood caused by contractions of the heart
d) process by which the bore of an arteriole becomes narrower, reducing blood flow to localised areas of the body
e) process by which the bore of an arteriole becomes wider, allowing increased blood flow to localised areas of the body
f) structure present in heart and veins that prevents backflow of blood
g) tissue in arteriole wall capable of contraction or relaxation, enabling blood flow to be controlled locally
h) tiny vessel that allows exchange of soluble substances between bloodstream and body tissues
i) small artery able to control local distribution of blood by vasoconstriction and vasodilation
j) large vessel that carries blood away from the heart and possesses a thick muscular wall
k) large vessel that carries blood back towards the heart and possesses a thin muscular wall and valves
l) name given to tissue fluid absorbed by and transported in lymph vessels
m) one of two short tubes by which lymph returns to the blood circulatory system
n) one of many tubes carrying lymph from tissues back towards the blood circulatory system
o) liquid bathing body cells that closely resembles blood plasma except that it contains little or no plasma protein
p) yellowish fluid component of blood that contains digested food, dissolved respiratory gases and plasma proteins
q) process by which blood is forced through capillaries causing much of its plasma to be squeezed out into the surrounding tissues

➡

Multiple Choice Test

Choose the ONE correct answer to each of the following multiple choice questions.

Questions 1, 2 and 3 refer to Figure 13.1, which shows the circulatory system.

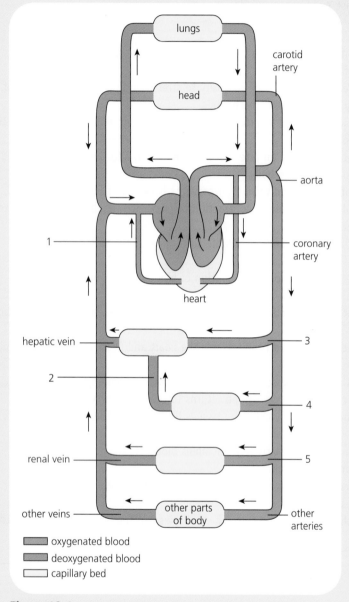

Figure 13.1

1 Which of the following statements is correct?
   A The jugular vein transports deoxygenated blood back towards the heart from the head.
   B The jugular vein transports oxygenated blood back towards the heart from the head.
   C The carotid vein transports deoxygenated blood back towards the heart from the head.
   D The carotid vein transports oxygenated blood back towards the heart from the head.

2 Which vessel begins in a capillary bed and ends in a capillary bed?
   A 1   B 2   C 3   D 4

3 Which vessel carries blood to the kidneys?
   A 2   B 3   C 4   D 5

4 On returning to the heart from the body tissues, blood is pumped to the lungs. Following oxygenation, blood is returned to the heart and then pumped out to the body. During this series of events, blood passes through four blood vessels in the order:

A pulmonary vein, vena cava, aorta, pulmonary artery.　　B pulmonary vein, aorta, vena cava, pulmonary artery.

C vena cava, pulmonary artery, pulmonary vein, aorta.　　D vena cava, pulmonary vein, pulmonary artery, aorta.

5 Which artery is immediately and directly affected during strangulation?

A aorta　　　　　　　　B carotid　　　　　　　　C coronary　　　　　　　　D pulmonary

Questions 6, 7 and 8 refer to the following list of features of certain blood vessels.

i) dissolved food and oxygen diffuse through their walls

ii) blood inside them is being transported back to the heart

iii) blood carried by them is at low pressure

iv) blood carried by them is at high pressure

v) their walls are only one cell thick

vi) their walls are thick and elastic

6 Which features BOTH apply to capillaries?

A i) and vi)　　　　　　B i) and v)　　　　　　C ii) and v)　　　　　　D ii) and vi)

7 Which features BOTH apply to arteries?

A i) and ii)　　　　　　B iii) and vi)　　　　　　C i) and iv)　　　　　　D iv) and vi)

8 Which features BOTH apply to veins?

A ii) and iii)　　　　　　B ii) and iv)　　　　　　C iv) and v)　　　　　　D iii) and vi)

Questions 9 and 10 refer to Figure 13.2, which shows part of an artery.

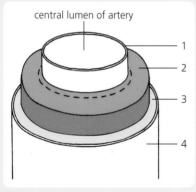

central lumen of artery

1
2
3
4

Figure 13.2

9 Which number indicates the endothelium?

A 1　　　　　　　　B 2　　　　　　　　C 3　　　　　　　　D 4

10 Which of the following layers BOTH contain elastic fibres?

A 1 and 2　　　　　　B 2 and 3　　　　　　C 1 and 4　　　　　　D 2 and 4

11 Table 13.1 compares an artery with a vein. Which row is correct?

	Feature	Artery	Vein
A	state of muscular wall	thinner	thicker
B	valves	present	absent
C	diameter of central lumen	narrower	wider
D	pressure of blood in vessel	lower	higher

Table 13.1

12 During vigorous activity, arterioles leading to capillary beds in skeletal muscles respond to nerve impulses by undergoing
  A vasoconstriction, which decreases blood flow to working muscles.
  B vasoconstriction, which increases blood flow to working muscles.
  C vasodilation, which decreases blood flow to working muscles.
  D vasodilation, which increases blood flow to working muscles.

Questions 13 and 14 refer to Figure 13.3, which shows the rate of blood flow in four parts of the body when the body is in different states of activity.

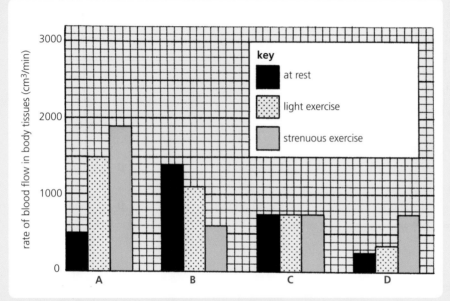

Figure 13.3

13 Which group of bars represents the small intestine?
14 Which group of bars represents the brain?
15 The data in Table 13.2 shows the results of an extensive survey of blood vessels.

Blood vessel	Mean external diameter	Mean internal diameter
capillary	10 µm	8 µm
artery	3.0 mm	1.3 mm
vein	3.6 mm	3.0 mm

Table 13.2

The mean internal diameter of a vein is greater than that of a capillary by a factor of
  A 36.0   B 37.5   C 360.0   D 375.0

16 As liquid is forced out of blood capillaries into the spaces between the cells in body tissues, it undergoes the process of
  A osmosis.   B diffusion.   C valve action.   D pressure filtration.

17 Compared with blood plasma, tissue fluid has a
  A higher water concentration and contains almost no protein.
  B lower water concentration and contains almost no protein.
  C higher water concentration and contains much more protein.
  D lower water concentration and contains much more protein.

Questions 18 and 19 refer to Figure 13.4, which shows the transport of materials through a capillary bed.

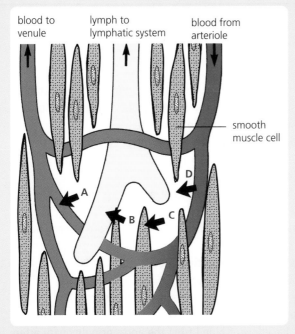

Figure 13.4

18 Which arrow represents osmotic return of tissue fluid?
19 Which arrow represents diffusion of oxygen into a respiring cell?
20 The flow of lymph in lymphatic vessels is brought about mainly by the
    A pumping action of lymph glands in the neck and groin.
    B osmotic pressure of fluid absorbed by body tissues.
    C pressure exerted by surrounding muscles on contraction.
    D two lymphatic ducts opening into veins from the arms.

14 Structure and function of the heart

Matching Test
Match the terms in list X with their descriptions in list Y.

list X
1 atrial systole
2 atrio-ventricular (AV)
3 atrio-ventricular node (AVN)
4 cardiac cycle
5 cardiac output
6 conducting fibres
7 diastole
8 electrocardiogram
9 heart rate
10 hypertension
11 semi-lunar (SL)
12 sino-atrial node (SAN)
13 sphygmomanometer
14 stethoscope
15 stroke volume
16 ventricular systole

list Y
a) pattern of contraction and relaxation shown by the heart during one heartbeat
b) phase of the cardiac cycle when muscle in the heart's upper chambers contracts
c) phase of the cardiac cycle when cardiac muscle is relaxed
d) phase of the cardiac cycle when muscle in the heart's lower chambers contracts
e) volume of blood pumped out of a ventricle per minute
f) number of heartbeats that occur per minute
g) quantity of blood expelled by each ventricle on contraction
h) instrument used to measure blood pressure
i) instrument that enables heart sounds to be heard when valves are forced shut during heartbeats
j) structures in ventricle walls that, on being stimulated, cause cardiac muscle cells to contract almost simultaneously
k) region of specialised tissue in the heart that picks up impulses from the SAN
l) prolonged elevation of blood pressure when at rest
m) type of valve that prevents backflow of blood from the aorta or pulmonary artery to a ventricle
n) type of valve that prevents backflow of blood from a ventricle to an atrium
o) pattern on an oscilloscope screen that represents the electrical activity involved in a heartbeat
p) group of autorhythmic cells (also known as the pacemaker) that sets the rate at which cardiac muscle cells contract

Multiple Choice Test
Choose the ONE correct answer to each of the following multiple choice questions.

Questions 1 and 2 refer to Figure 14.1, which shows the human heart.

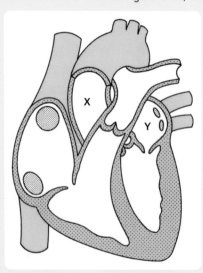

Figure 14.1

1 Vessel X carries
  A oxygenated blood from the heart.
  B deoxygenated blood to the lungs.
  C oxygenated blood from the lungs.
  D deoxygenated blood to the heart.

2 Chamber Y
  A pumps blood to the left atrium.
  B pumps blood to the right ventricle.
  C collects blood from the lungs.
  D collects blood from all parts of the body.

3 Which of the following equations is correct? (HR = heart rate, SV = stroke volume and CO = cardiac output)

A $SV = \dfrac{HR}{CO}$ B $CO = \dfrac{HR}{SV}$ C $SV = HR \times CO$ D $CO = HR \times SV$

Questions 4 and 5 refer to Table 14.1, which shows the effect of exercise on cardiac output.

State of body	Heart rate (beats/min)	Stroke volume (ml)	Cardiac output by each ventricle (l/min)
at rest	70	60	4.2
during exercise	110	70	box X
during vigorous exercise	160	box Y	12.0

Table 14.1

4 The cardiac output result that should have been entered in box X is

A 1.6   B 7.7   C 7.8   D 7700

5 The stroke volume that should have been entered in box Y is

A 13.3   B 75   C 80   D 1920

6 The graph in Figure 14.2 records the pulse rate of a woman taken at one-minute intervals before, during and after a period of vigorous exercise.

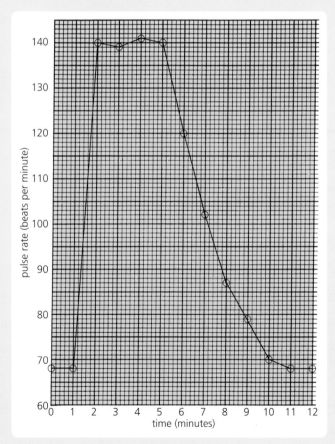

Figure 14.2

During which time interval in minutes did her pulse rate decrease at the fastest rate?

A 5–6   B 6–7   C 7–8   D 8–9

7 Figure 14.3 shows three stages in the cardiac cycle.

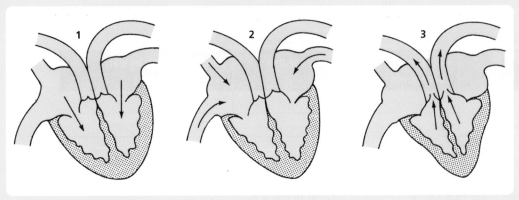

Figure 14.3

Which of the following is the correct sequence?

A 2, 3, 1   B 1, 2, 3   C 2, 1, 3   D 3, 1, 2

Questions 8 and 9 refer to Figure 14.4, which represents one cardiac cycle lasting 0.8 s, and to the possible answers that follow it.

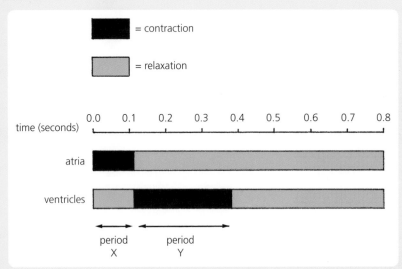

Figure 14.4

A atrial diastole and ventricular systole   B atrial diastole and ventricular diastole

C atrial systole and ventricular systole   D atrial systole and ventricular diastole

8 Which answer describes the events that occur during period X?

9 Which answer describes the events that occur during period Y?

10 Which of the following blood vessels BOTH possess semi-lunar valves?

    A vena cava and aorta   **B** pulmonary vein and vena cava

    C aorta and pulmonary artery   **D** pulmonary artery and pulmonary vein

Questions 11, 12, 13 and 14 refer to Figure 14.5, which shows pressure changes affecting the aorta, the left atrium and the left ventricle.

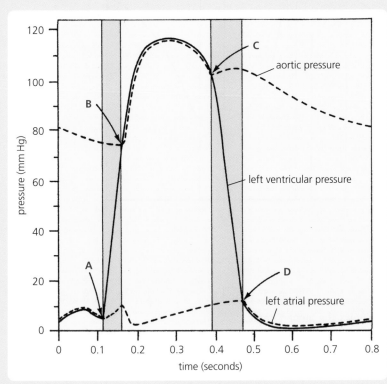

Figure 14.5

11 Which letter indicates the point at which the semi-lunar valve opens?

12 Which letter indicates the point at which the atrio-ventricular valve opens?

13 Which letter points to the start of ventricular systole?

14 Which letter indicates the point at which the semi-lunar valve closes?

15 'Lubb', the first heart sound in the cardiac cycle, is caused by the

    A opening of the AV valve.   **B** opening of the SL valve.

    C closing of the AV valve.   **D** closing of the SL valve.

16 Figure 14.6 refers to the volume of blood present in the ventricles during two complete cardiac cycles.

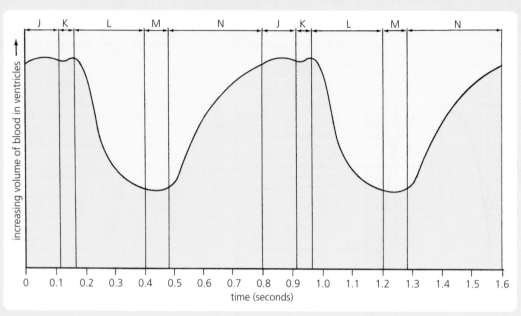

Figure 14.6

Ventricular diastole occurs during period

A J, K and L.   **B** K, L and M.   **C** L, M and N.   **D** M, N and J.

Questions 17 and 18 refer to Figure 14.7, which shows the human heart.

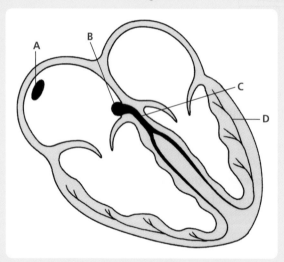

Figure 14.7

17 Which lettered structure is a region of specialised tissue able to initiate the electrical impulses that make cardiac muscle cells contract at a certain rate?

18 Which structure is the atrio-ventricular node?

19 A decrease in heart rate occurs when the number of impulses conducted to the sino-atrial node by the

  A sympathetic nerve decreases.   **B** sympathetic nerve increases.

  C parasympathetic nerve decreases.   **D** parasympathetic nerve increases.

Questions 20 and 21 refer to Figure 14.8, which shows a small part of a normal electrocardiogram.

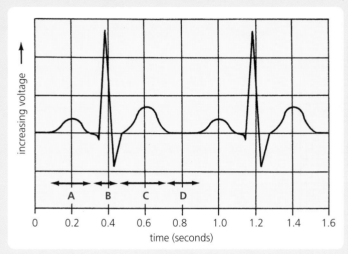

Figure 14.8

20 Which region represents a wave of excitation passing through the ventricles?
21 Which region corresponds to the wave of electrical excitation spreading over the atria from the SAN (sino-atrial node)?
22 During fibrillation, individual muscle fibres contract in a disorderly sequence and regular coordinated beating of the affected heart chambers fails to occur. Which of the electrocardiograms in Figure 14.9 shows ventricular fibrillation?

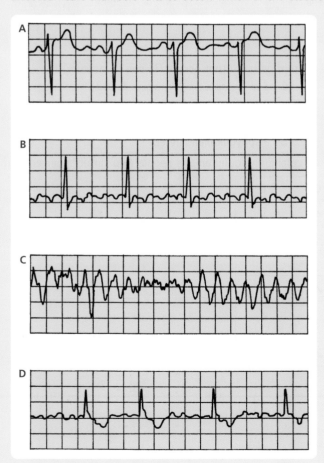

Figure 14.9

23 The graph in Figure 14.10 shows the change in blood pressure that occurs as blood passes through the circulatory system.

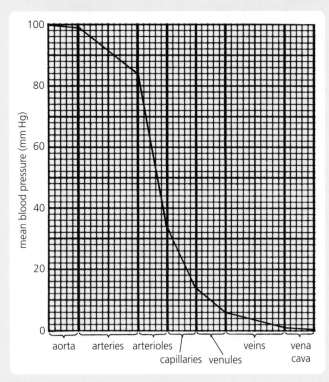

Figure 14.10

From entering the arterioles to leaving the venules, the drop in pressure that occurs (in mm of mercury) is

A 50   B 78   C 84   D 93

24 Blood pressure is measured using

A a stethoscope.   B a phonocardiogram.
C an electrocardiogram.   D a sphygmomanometer.

25 Which column in Table 14.2 indicates features that are all commonly found among people who suffer hypertension?

Feature	A	B	C	D
being underweight	✓		✓	
being overweight		✓		✓
rarely taking exercise	✓	✓		
exercising regularly			✓	✓
frequently drinking alcohol to excess		✓		✓
rarely drinking alcohol	✓		✓	

Table 14.2

15 Pathology of cardiovascular disease

Matching Test Part 1
Match the terms in list X with their descriptions in list Y.

list X
1 angina
2 atheroma
3 atherosclerosis
4 calcium
5 cardiovascular
6 cholesterol
7 endothelium
8 familial hypercholesterolaemia (FH)
9 high-density lipoprotein
10 low-density lipoprotein
11 peripheral vascular disease
12 statins

list Y
a) type of molecule that transports cholesterol to body cells
b) type of molecule that transports cholesterol from body cells to the liver
c) inner lining in the wall of an artery
d) atherosclerosis of arteries other than aorta, coronary and carotid arteries
e) type of coronary heart disease characterised by pain in the chest, neck and left arm
f) plaque deposited in an artery wall and composed of cholesterol, fibrous material and calcium
g) chemical element that hardens atheromas
h) term referring to the heart and blood transport systems
i) formation of atheromas beneath the endothelium in the wall of an artery
j) basic component of cell membrane and precursor for the synthesis of steroids
k) drugs that reduce blood cholesterol by inhibiting its synthesis in liver cells
l) inherited disorder characterised by very high LDL-cholesterol levels in the bloodstream

Matching Test Part 2
Match the terms in list X with their descriptions in list Y.

list X
1 clotting factors
2 embolus
3 fibrin
4 fibrinogen
5 myocardial infarction
6 prothrombin
7 pulmonary embolism
8 thrombin
9 thrombosis
10 thrombus

list Y
a) enzyme that causes fibrinogen to become threads of fibrin
b) inactive form of the enzyme thrombin present in blood plasma
c) soluble plasma protein that is converted to insoluble fibrin by thrombin
d) chemical in the form of a network of threads to which platelets adhere, forming a blood clot
e) blood clot formed during thrombosis
f) a thrombus that has broken loose and is able to travel through the bloodstream until it blocks a narrow vessel
g) formation of a blood clot
h) chemicals released by damaged cells that activate a series of reactions leading to blood clotting
i) heart attack resulting from a thrombus blocking a coronary artery
j) blockage of a small branch of the pulmonary artery by an embolus

Multiple Choice Test

Choose the ONE correct answer to each of the following multiple choice questions.

1   Figure 15.1 shows a cross section of an artery of a sufferer of atherosclerosis. Which letter indicates an atheroma?

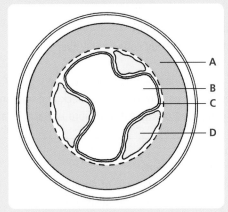

— A
— B
— C
— D

Figure 15.1

2   An atheroma consists mainly of
   A   cholesterol and fibrous material hardened by calcium.
   B   cholesterol and fibrous material hardened by iron.
   C   fibrous material and protein hardened by calcium.
   D   fibrous material and protein hardened by iron.

3   Which row in Table 15.1 correctly represents the effects brought about by the development of a large atheroma?

	Artery's elasticity	Blood pressure	Diameter of artery's lumen
A	↑	↑	↓
B	↓	↓	↑
C	↑	↓	↑
D	↓	↑	↓

(↑ = increased, ↓ = decreased)

Table 15.1

4   Which of the following is NOT caused by atherosclerosis?
   A   stroke   B   angina
   C   haemophilia   D   peripheral vascular disease

5 Which of the diagrams in Figure 15.2 correctly represents the chemical reactions that lead to the clotting of blood?

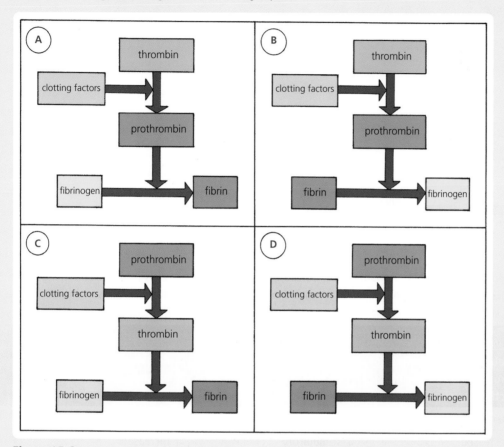

Figure 15.2

6 The boxes listed in Figure 15.3 contain some of the events that lead to a 'heart attack'.

1. blood clot breaks loose and passes along blood vessel

2. region of heart muscle suffers a myocardial infarction

3. atheroma ruptures, damaging the endothelium of coronary artery

4. narrow branch of coronary artery becomes blocked

5. thrombus forms on inner surface of coronary artery

Figure 15.3

The correct sequence of these events is

A 3, 5, 1, 2, 4   B 3, 5, 1, 4, 2   C 5, 3, 1, 2, 4   D 5, 3, 1, 4, 2

7 Under normal circumstances, once a blood clot has served its purpose, the fibrin it contains is broken down as indicated in Figure 15.4.

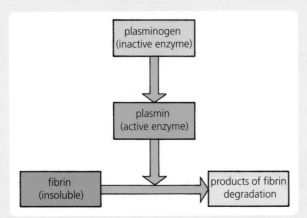

Figure 15.4

Thrombolysis is the process by which an unwanted clot blocking a narrow blood vessel is broken down using medication. Table 15.2 gives the laboratory findings for four medications. Which could be used as a thrombolytic drug?

Medication	Laboratory finding		
	Relative time required for prothrombin to become active	Effect on blood-clotting factors	Relative time required for plasminogen to become plasmin
A	increased	stimulatory	decreased
B	decreased	stimulatory	increased
C	increased	inhibitory	decreased
D	decreased	inhibitory	increased

Table 15.2

8 Which of the following could be affected by peripheral vascular disease?

A   aorta   B   leg artery   C   carotid artery   D   coronary artery

9 Which of the diagrams in Figure 15.5 shows the route that would be taken by an embolus that begins as a deep vein thrombus and ends up causing a pulmonary embolism?

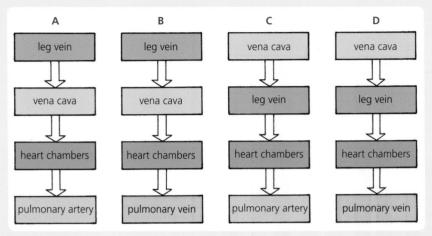

Figure 15.5

10 Which of the following statements about cholesterol is NOT correct?
   A It is a basic component of the cell membrane.
   B It is a precursor for the synthesis of steroids.
   C Its level in the bloodstream is increased by a diet rich in saturated fat.
   D It is synthesised from unsaturated fat in liver cells only.

11 Which of the following statements is correct? (Note: HDL = high density lipoprotein and LDL = low density lipoprotein)
   A HDL transports cholesterol from body cells to the liver for elimination.
   B HDL transports cholesterol from the liver to body cells for storage.
   C LDL transports cholesterol from body cells to the liver for storage.
   D LDL transports cholesterol from the liver to body cells for elimination.

12 Table 15.3 refers to the relative ratio of HDL and LDL in the blood and two related effects. Which row is correct?

	Relative ratio of HDL to LDL in blood	Effect 1	Effect 2
		Relative level of cholesterol in blood	Chance of atherosclerosis
A	higher	higher	increased
B	lower	lower	increased
C	higher	lower	reduced
D	lower	higher	reduced

Table 15.3

13 Figure 15.6 shows the percentage decrease in death rate from coronary heart disease (CHD) between 1996 and 2006 in some European countries.

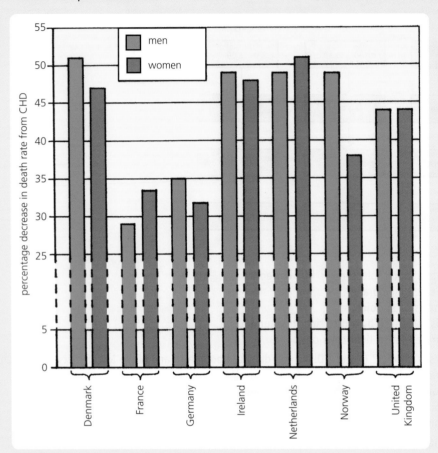

Figure 15.6

Which of the following conclusions can be correctly drawn from the data?

A Compared with women, a greater percentage decrease in deaths from CHD was found among men in all of these countries.

B Compared with women in Ireland, women in Denmark showed a greater percentage decrease in deaths from CHD.

C Compared with men in Norway, men in the United Kingdom showed the same percentage decrease in deaths from CHD.

D Compared with women in France, women in Germany showed a smaller percentage decrease in deaths from CHD.

14 Which row in Table 15.4 gives the correct answers to blanks 1, 2 and 3 in the following passage?

When body cells have sufficient cholesterol for their requirements, synthesis of further LDL _____1_____ at cell surfaces is inhibited by _____2_____ feedback control. Therefore, molecules of _____3_____ carrying cholesterol circulate in the bloodstream and may deposit their cholesterol in the walls of arteries.

	Blank 1	Blank 2	Blank 3
A	molecules	positive	LDL
B	receptors	negative	LDL
C	molecules	positive	HDL
D	receptors	negative	HDL

Table 15.4

15 Statins reduce blood cholesterol level by

   A inhibiting the synthesis of cholesterol in liver cells.

   B increasing the level of HDL in the bloodstream.

   C converting saturated fats to unsaturated fats.

   D promoting the breakdown of LDL molecules in the bloodstream.

Questions 16 and 17 refer to Table 15.5, which gives the results of a clinical trial investigating the action of a type of statin.

Condition	Percentage of test group affected after 5 years		Percentage risk reduction
	Control group	**Statin group**	
major cerebrovascular event	2.5	2.0	box X
major coronary event	5.6	box Y	25
all-cause mortality	6.0	5.4	10

Table 15.5

16 The correct value for box X is

   A 0.5   **B** 2.0   **C** 20.0   **D** 25.0

17 The correct value for box Y is

   A 1.4   **B** 4.2   **C** 7.0   **D** 9.8

Questions 18 and 19 refer to Figure 15.7, which shows a pedigree for familial hypercholesterolaemia (FH) where H (the mutant allele for this inherited condition) is dominant to h (the allele for the unaffected state).

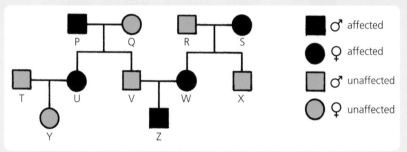

Figure 15.7

18 Which line in Table 15.6 correctly identifies the genotypes of persons P and W?

	P	W
A	Hh	Hh
B	HH	Hh
C	Hh	HH
D	HH	HH

Table 15.6

19 If persons V and W have a second child, the chance of this person being affected by FH is

   A 1 in 1.   **B** 1 in 2.   **C** 1 in 3.   **D** 1 in 4.

20 Which of the following features is NOT typical of an untreated sufferer of familial hypercholesterolaemia (FH)?

   A reduction in relative number of LDL receptors present in cell membranes

   B presence in bloodstream of many LDL molecules unable to release their cholesterol to cells

   C occurrence of serious cardiovascular problems at a very early age

   D presence in bloodstream of high concentration of HDL molecules carrying cholesterol

16 Blood glucose levels and obesity

Matching Test
Match the terms in list X with their descriptions in list Y.

list X
1 adrenaline (epinephrine)
2 body mass index (BMI)
3 diabetes
4 glucagon
5 glucose tolerance test
6 glycogen
7 haemorrhage
8 insulin
9 microvascular disease
10 negative feedback control
11 obesity
12 urine test

list Y
a) hormone that activates the enzyme that catalyses the conversion of glucose to glycogen
b) hormone that activates the enzyme that catalyses the conversion of glycogen to glucose
c) hormone released during 'fight or flight' that stimulates glucagon secretion
d) release of blood by a burst blood vessel into surrounding tissues
e) storage carbohydrate composed of glucose molecules joined together
f) condition where the untreated sufferer is unable to control their blood glucose concentration
g) method used to diagnose diabetes based on the body's ability to deal with ingested glucose
h) method used as an indicator of diabetes
i) condition characterised by accumulation of excess body fat in relation to lean muscle tissue
j) indicator of state of health based on the relationship between a person's weight and height
k) disorder caused by small vessels becoming damaged and leaking blood into surrounding tissues
l) self-regulating mechanisms that maintain the body's steady state by counteracting deviations from the norm

Multiple Choice Test
Choose the ONE correct answer to each of the following multiple choice questions.

1 When the glucose concentration of the blood becomes chronically elevated, the events listed in Figure 16.1 may occur.

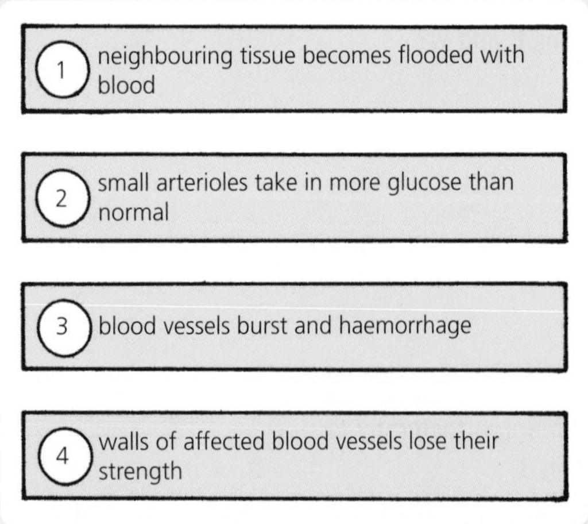

1 neighbouring tissue becomes flooded with blood

2 small arterioles take in more glucose than normal

3 blood vessels burst and haemorrhage

4 walls of affected blood vessels lose their strength

Figure 16.1

The correct sequence of events is
A 2, 4, 3, 1   B 2, 4, 1, 3   C 4, 2, 3, 1   D 4, 2, 1, 3

2   Which of the following are ALL examples of tissues commonly affected by microvascular disease in people whose blood glucose concentration is chronically elevated?

A   lung, kidney and retina

B   retina, peripheral nerves and brain

C   brain, lung and kidney

D   kidney, retina and peripheral nerves

3   Negative feedback control involves the following four stages:

W   effectors bring about corrective responses

X   a receptor detects a change in the internal environment

Y   deviation from the norm is counteracted

Z   nerve or hormonal messages are sent to effectors

The order in which these stages occur is

A   X, Z, W, Y.   B   X, W, Z, Y.   C   X, Z, Y, W.   D   Z, X, W, Y.

4   Figure 16.2 shows a simplified version of the homeostatic control of blood sugar level in the human body.

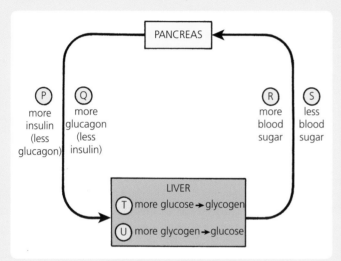

Figure 16.2

Which of the following responses would occur in a human body as a direct result of eating a meal rich in carbohydrate?

A   S, Q, U   B   R, Q, T   C   S, P, U   D   R, P, T

Questions 5 and 6 refer to Figure 16.3, which shows the blood sugar concentration of a person who has consumed 50 g of glucose at the time indicated.

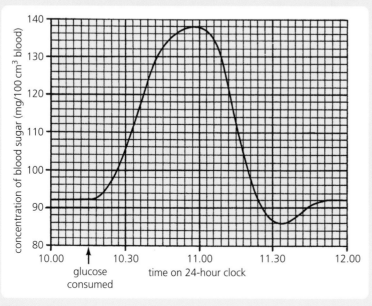

Figure 16.3

5 The maximum level of blood sugar expressed as a multiple of the initial value is

A 0.5   B 0.67   C 1.5   D 12 696

6 Compared with their concentrations at 10.00 hours, at which of the following two times could increased concentrations of insulin and glucagon be present?

	Extra insulin	Extra glucagon
A	10.30	11.00
B	10.30	11.30
C	11.00	10.30
D	11.30	10.30

Table 16.1

7 Which of the glucose tolerance curves in the graph in Figure 16.4 indicates a mild form of type 2 diabetes that may respond to control by a diet and not need medication?

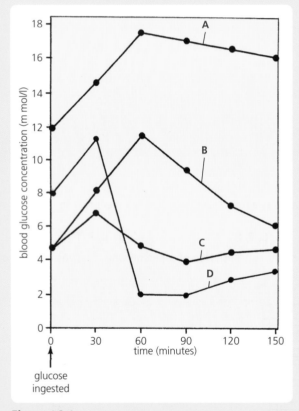

Figure 16.4

8 During the 'fight or flight' response, glucose levels are raised by adrenaline (epinephrine)

A inhibiting the secretion of glucagon and insulin.

B stimulating the secretion of insulin and glucagon.

C inhibiting glucagon secretion and stimulating insulin secretion.

D stimulating glucagon secretion and inhibiting insulin secretion.

Questions 9 and 10 refer to Figure 16.5, where the four lettered structures represent glands that produce hormones.

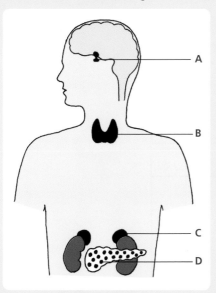

Figure 16.5

9 Which gland is the site of glucagon production?

10 Which of these glands makes adrenaline (epinephrine)?

11 Individuals with type 2 diabetes have a relatively low number of insulin receptors on the surfaces of their liver cells. This results in the normal conversion of
  A glucose to glycogen being prevented.
  B glycogen to glucose being prevented.
  C glucose to glycogen being promoted.
  D glycogen to glucose being promoted.

12 Table 16.2 compares types 1 and 2 diabetes. Which row in the table is NOT correct?

		Type 1	Type 2
A	typical body mass of sufferer	normal or underweight	overweight or obese
B	stage of life at which condition normally first occurs	childhood or early teens	adulthood
C	ability of pancreatic cells to produce insulin	present	absent
D	percentage of all cases	5–10	90–95

Table 16.2

Questions 13, 14 and 15 refer to the chart in Figure 16.6, which can be used to find out if a person's body weight is correct for their height.

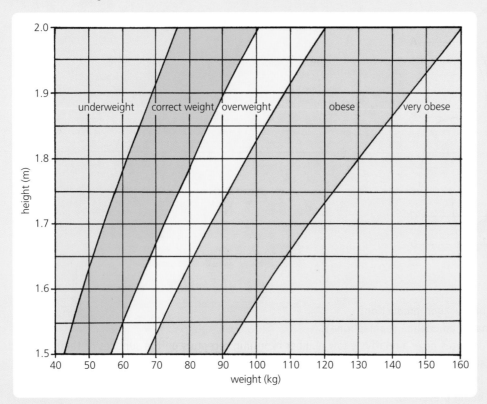

Figure 16.6

13 Which person described in Table 16.3 is obese?

Person	Height (m)	Weight (kg)
A	1.70	70
B	1.75	95
C	1.80	70
D	1.85	95

Table 16.3

14 If a person is 1.7 m in height and overweight (but not obese), then their weight (in kg) could be

A 45   B 60   C 75   D 90

15 What is the minimum correct weight (in kg) for a person of height 1.85 m?

A 40   B 65   C 75   D 86

16 A person's body mass index (BMI) is calculated using the formula:

$$BMI = \frac{\text{body mass (kg)}}{\text{height (m)}^2}$$

Table 16.4 gives a classification of BMI values.

BMI value	Opinion of medical experts	Risk of health problems associated with obesity
20.0–25.0	ideal for height	average
25.1–30.0	overweight	increased
30.1–40.0	obese	greatly increased
over 40.0	very obese	very greatly increased

Table 16.4

Person X has a body mass of 104 kg and is 180 cm in height. Person Y has a body mass of 76 kg and is 170 cm in height. Which row in Table 16.5 correctly identifies their respective level of risk of health problems associated with obesity?

	Risk of health problems associated with obesity	
	Person X	Person Y
A	increased	increased
B	greatly increased	average
C	increased	average
D	greatly increased	increased

Table 16.5

Questions 17 and 18 refer to Tables 16.6 and 16.7. Table 16.6 shows a set of results for five people aged 18 years using skinfold callipers. Table 16.7 shows percentage fat content of the body for different skinfold thicknesses.

Location tested	Mean skinfold thickness (mm) using skinfold callipers				
	Marion	Callum	Iyesha	Steven	Kirsty
side of waist	21	12	18	25	15
back of upper arm	19	10	15	22	13
front of upper arm	10	6	8	12	7
back below shoulder blade	20	12	19	21	15

Table 16.6

Total skinfold thickness for the four locations (mm)	Fat content of body (%)	
	Women aged 16–20	Men aged 16–20
20	14.0	8.0
30	19.5	13.0
40	23.5	16.5
50	26.5	19.0
60	29.0	21.0
70	31.0	23.0
80	33.0	25.0

Table 16.7

17 What is the percentage fat content of Kirsty's body?

  A 19.0 **B** 26.5 **C** 29.0 **D** 50.0

18 Whose body contains 25 per cent fat?

  A Steven **B** Iyesha **C** Callum **D** Marion

19 Which tube in the heart shown in Figure 16.7 correctly represents the result of a successful coronary bypass operation?

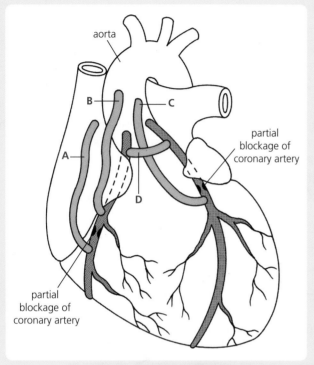

Figure 16.7

20 How many days would it take a person to lose 4 kg of body fat if their total energy output exceeds their total energy input by 784 kJ/day? (Note: 1 kg of body fat contains 29.4 MJ of energy)

  A 15 **B** 150 **C** 1500 **D** 150 000

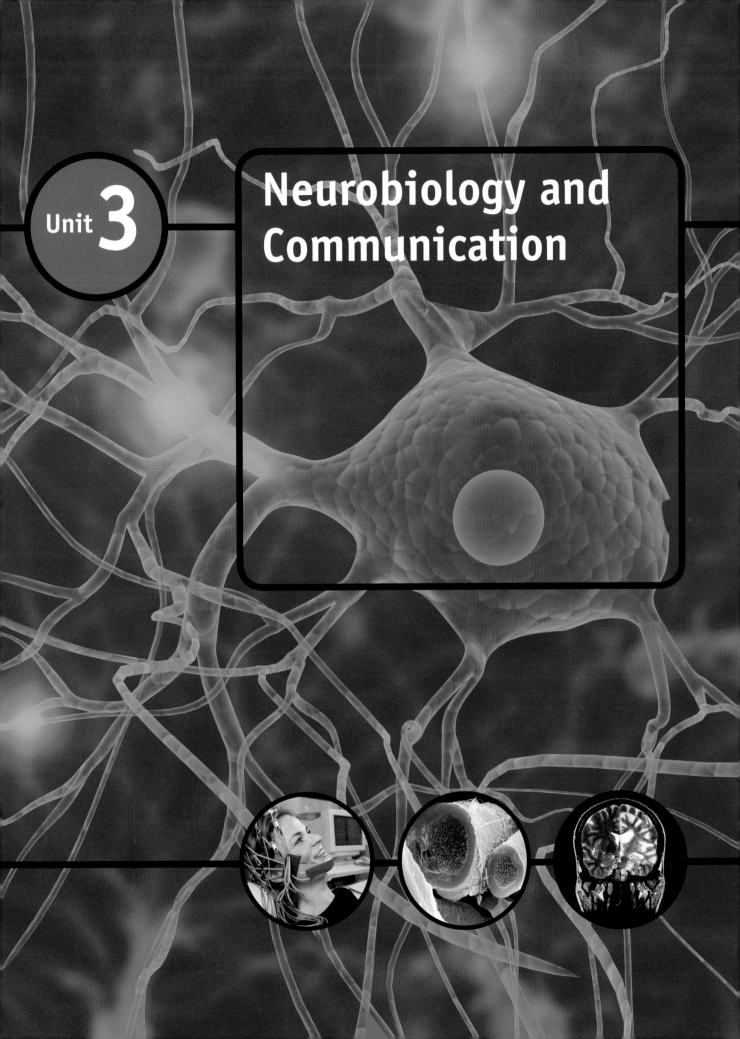

Unit 3

Neurobiology and Communication

17 Divisions of the nervous system and parts of the brain

Matching Test Part 1

Match the terms in list X with their descriptions in list Y.

list X
1 antagonistic
2 autonomic nervous system (ANS)
3 central nervous system (CNS)
4 parasympathetic
5 peripheral nervous system (PNS)
6 somatic nervous system (SNS)
7 sympathetic

list Y
a) subdivision of the 'functional' nervous system that regulates the internal environment and controls organs by involuntary means
b) subdivision of the 'functional' nervous system that controls skeletal muscles and allows voluntary action
c) subdivision of the 'structural' nervous system comprised of cranial and spinal nerves
d) subdivision of the 'structural' nervous system containing brain and spinal cord
e) subdivision of the ANS that arouses the body in preparation for 'fight or flight'
f) subdivision of the ANS that prepares the body for 'rest and digest'
g) describing two systems that work by mutually opposing one another

Matching Test Part 2

Match the terms in list X with their descriptions in list Y.

list X
1 association area
2 central core
3 cerebellum
4 cerebral cortex
5 cerebrum
6 corpus callosum
7 limbic system
8 medulla
9 motor area
10 sensory area

list Y
a) outer convoluted layer of cerebrum that is the centre of conscious thought
b) region of the cerebrum that receives information from the body's receptors
c) region of the cerebrum that sends impulses to effectors
d) region of the brain that contains the medulla and the cerebellum
e) region of the brain that regulates breathing, heart rate, arousal and sleep
f) region of the brain that is responsible for controlling balance, posture and movement
g) region of the cerebrum that analyses and interprets information from sensory area
h) composite region of the brain that processes information for memories and influences emotional and motivational states
i) largest part of the brain; divided into two hemispheres
j) large bundle of nerve fibres connecting cerebral hemispheres

Multiple Choice Test

Choose the ONE correct answer to each of the following multiple choice questions.

1 Figure 17.1 shows an incomplete classification of the nervous system based on function.

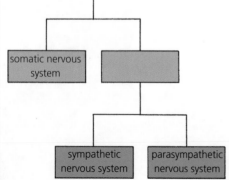

The blank box should contain the words
A central nervous system.
B peripheral nervous system.
C voluntary nervous system.
D autonomic nervous system.

Figure 17.1

2 Which of the diagrams in Figure 17.2 correctly represents the flow of information through the nervous system?

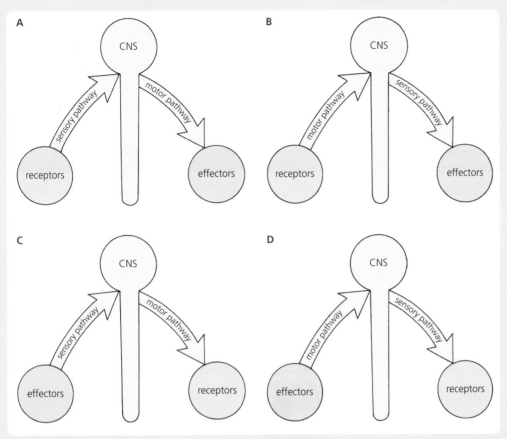

Figure 17.2

3 Which row in Table 17.1 correctly describes the situation that occurs when the type of reflex action called limb withdrawal takes place?

	Type of nervous system involved	Type of response
A	somatic	voluntary
B	autonomic	voluntary
C	somatic	involuntary
D	autonomic	involuntary

Table 17.1

4 Which of the following differences between the sympathetic and parasympathetic nervous systems is NOT correct?

	Sympathetic	Parasympathetic
A	prepares the body for 'fight or flight'	prepares the body for 'rest and digest'
B	involves expenditure of energy	promotes conservation of energy
C	promotes hyperactivity	promotes hypoactivity
D	operates under voluntary control	operates under involuntary control

Table 17.2

5 Stimulation of the parasympathetic system leads to an increase in

A dilation of bronchioles.
B rate of secretion of epinephrine.
C rate of peristalsis.
D cardiac output of blood.

6 Stimulation of the sympathetic nervous system does NOT lead to an increase in

A blood flow in coronary arteries.
B volume of blood flowing to the skin.
C volume of air entering the lungs.
D rate of heart beat.

7 Table 17.3 gives examples of the effects of the autonomic nervous system on certain parts of the body. Which entry is NOT correct?

	Branch of autonomic nervous system	Part of body innervated	Effect produced
A	parasympathetic	muscle of heart wall	rate of contraction decreased
B	sympathetic	sweat glands	secretion stimulated
C	parasympathetic	muscle of bronchi	relaxation promoted
D	sympathetic	glands in intestinal wall	secretion inhibited

Table 17.3

8 Figure 17.3 shows in a simplified way how the activity of two sets of antagonistic muscles in the iris of the eye controls the size of the pupil.

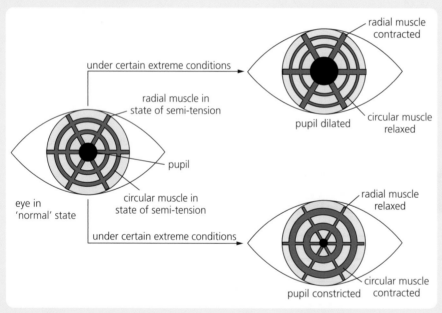

Figure 17.3

Which of the following would occur during a 'fight or flight' crisis?

	Part of autonomic nervous system involved	Response by radial muscles of iris	Response by circular muscles of iris	Resultant state of pupil
A	sympathetic	contraction	relaxation	dilated
B	parasympathetic	relaxation	contraction	constricted
C	sympathetic	relaxation	contraction	constricted
D	parasympathetic	contraction	relaxation	dilated

Table 17.4

9 Which of the following would result if a person were made to re-breathe their own air?

	CO_2 concentration of blood	Number of nerve impulses sent by chemoreceptors to medulla	Rate and depth of breathing
A	↓	↓	↓
B	↑	↓	↑
C	↓	↑	↓
D	↑	↑	↑

(↑ = increase, ↓ = decrease)

Table 17.5

10 Figure 17.4 shows a cross section of the human brain.

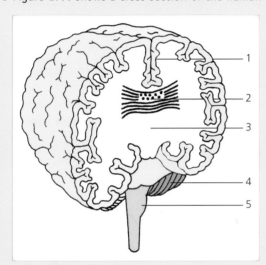

Figure 17.4

Which structures are known collectively as the central core?

A   1 and 2   B   2 and 4   C   3 and 5   D   4 and 5

11 Which of the following is NOT a function of the limbic system?
A   controlling balance   B   regulating emotional states
C   influencing biological motivation   D   processing information to form long-term memories

12 Which row in Table 17.6 is correct?

	Medulla	Cerebellum
A	regulates posture	controls sleep
B	regulates heart rate	controls sleep
C	regulates posture	controls muscular coordination
D	regulates heart rate	controls muscular coordination

Table 17.6

13 The corpus callosum allows rapid transfer of information from

  A one cerebral hemisphere to the other.   B the spinal cord to the cerebral hemispheres.

  C the cerebrum to the cerebellum.   D the spinal cord to the cerebellum.

14 Which row in Table 17.7 is correct?

		Controlling centre	
		left cerebral hemisphere	right cerebral hemisphere
A	region of human body controlled	lower half	upper half
B		upper half	lower half
C		right side	left side
D		left side	right side

Table 17.7

15 Information from the body's receptors passes through the discrete areas of the cerebrum in the order

  A motor, sensory, association.   B sensory, association, motor.

  C motor, association, sensory.   D sensory, motor, association.

Questions 16, 17 and 18 refer to Figures 17.5 and 17.6. Figure 17.5 indicates the share of the brain's somatosensory area allocated to each body part; Figure 17.6 shows an imaginary human figure ('sensory homunculus') whose body parts have been drawn in proportion to their sensitivity as opposed to their actual size.

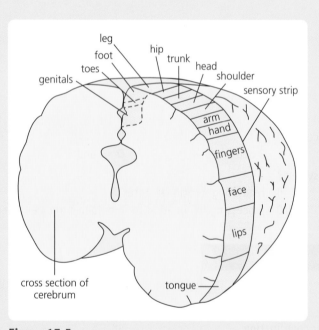

Figure 17.5

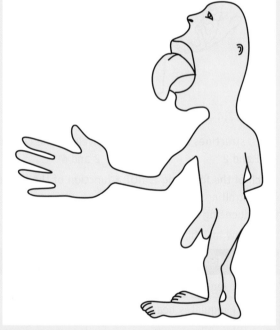

Figure 17.6

16 Which of the following is a large part of a normal human body yet is represented by a relatively small part of the cerebrum's somatosensory region?

  A tongue   B trunk   C face   D hand

17 Which of the following body parts contains most sensory receptors relative to its actual size?

  A leg   B shoulder   C hip   D tongue

18 Which of the following structures would have fewest sensory nerve endings in relation to their actual size?

  A lips   B arms   C fingers   D genitals

19 The diagrams in Figure 17.7 show EEGs (electroencephalograms), which are records of electrical activity generated by the brain. Which pattern in the second diagram best represents an EEG of a person in a coma?

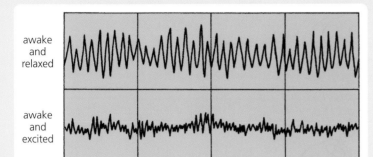

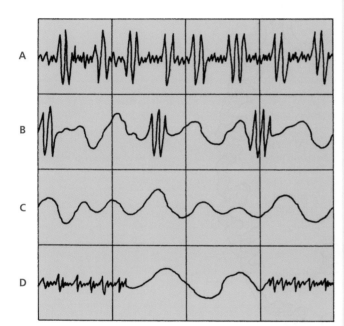

Figure 17.7

Questions 20, 21 and 22 depend on you understanding the information contained in Figure 17.8, which shows the cerebral hemispheres.

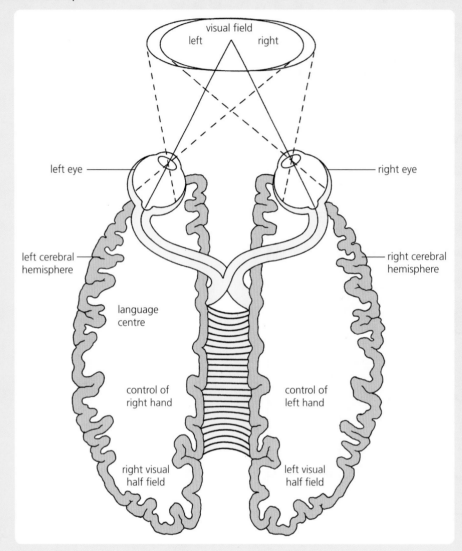

Figure 17.8

Questions 20 and 21 refer to the split-brain patient shown in Figure 17.9.

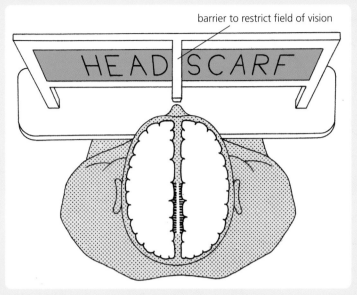

barrier to restrict field of vision

HEAD SCARF

Figure 17.9

20 When asked to say what he sees, the person would answer
   A head.   **B** scarf.   **C** scarfhead.   **D** headscarf.

21 When asked to turn around and point with his left hand to a card showing what he has just seen on the screen, he would choose
   A head.   **B** scarf.   **C** scarfhead.   **D** headscarf.

22 The same split-brain patient is found to be unable to state the name of an easily identified object such as a fork when it is placed in his left hand out of sight. This is because
   A the brain's language centre is damaged and unable to control the formation of speech.
   B the patient is right-handed and unable to identify familiar objects with his left hand.
   C the left hand is controlled by the right cerebral hemisphere, which has no access to the language centre.
   D the region of the brain that controls movement of the left hand is no longer able to operate.

Questions 23 and 24 refer to Figure 17.10, which shows a map of the brain's language areas.

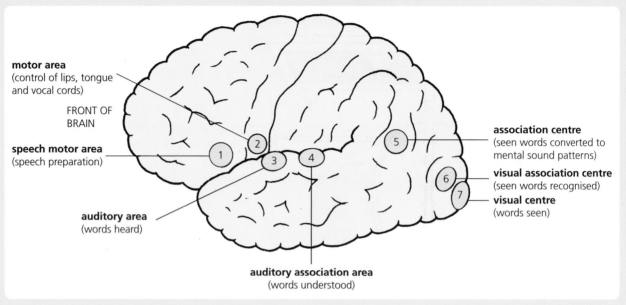

motor area
(control of lips, tongue and vocal cords)

FRONT OF BRAIN

speech motor area
(speech preparation)

auditory area
(words heard)

auditory association area
(words understood)

association centre
(seen words converted to mental sound patterns)

visual association centre
(seen words recognised)

visual centre
(words seen)

Figure 17.10

23 If you were to read this question to a deaf person (who is able to lip-read) by moving your lips silently in an exaggerated way, then a high level of activity would be registered in the language areas of YOUR brain in the order

A 7, 6, 5, 4, 1, 2

B 6, 7, 5, 4, 2, 1

C 7, 6, 4, 3, 2, 1

D 6, 7, 5, 4, 3, 2

24 If you listened carefully to a message on a pair of headphones and then repeated it out loud, a high level of activity would be registered in the language areas of your brain in the order

A 4, 3, 2, 1

B 3, 4, 2, 1

C 4, 3, 1, 2

D 3, 4, 1, 2

25 The region NOT duplicated on both sides of the brain is the

A auditory area.

B speech motor area.

C visual association centre.

D auditory association centre.

Matching Test Part 1

Match the terms in list X with their descriptions in list Y.

list X
1 binocular disparity
2 perception
3 perceptual constancy
4 perceptual set
5 segregation
6 superimposition
7 visual cue

list Y
a) process by which the brain analyses and makes sense of incoming sensory information
b) perceptual organisation of stimuli into coherent discrete images and patterns
c) indicator that helps the brain to perceive the distance of an object from the eye
d) partial blocking of the view of one object by another
e) difference between the images of the same object viewed simultaneously by both eyes
f) maintenance of stable perception of surroundings despite the movement of objects towards or away from the eyes
g) tendency to perceive certain aspects of available sensory information and ignore others

Matching Test Part 2

Match the terms in list X with their descriptions in list Y.

list X
1 chunking
2 contextual cue
3 elaboration
4 encoding
5 episodic memory
6 long-term memory (LTM)
7 organisation
8 procedural memory
9 rehearsal
10 retrieval
11 semantic memory
12 sensory memory (SM)
13 short-term memory (STM)
14 storage

list Y
a) conversion of information into a form that can be stored and retrieved
b) retention of information in the memory
c) recovery of material that has been stored in the memory
d) practice of material in order to improve its chance of being encoded and retained in the LTM
e) creating a meaningful unit of information from several smaller ones
f) the recall of personal facts and experiences
g) the recall of impersonal facts and concepts
h) memory system with limited capacity that retains information for periods of about 30 seconds
i) memory system with unlimited capacity that retains information for long periods
j) memory system that retains sensory images for only a second or two but sends a relevant selection to the STM
k) type of memory that contains information needed to perform a motor or mental skill
l) grouping of items into categories for ease of storage in the memory
m) consideration of detailed meaning and features of an item in order to make it more memorable
n) reminder that aids retrieval of information from the LTM

➡

Multiple Choice Test

Choose the ONE correct answer to each of the following multiple choice questions.

1   Which of the following factors determines how the visual system organises the objects shown in Figure 18.1 into groups?

A   closure   B   proximity   C   orientation   D   similarity

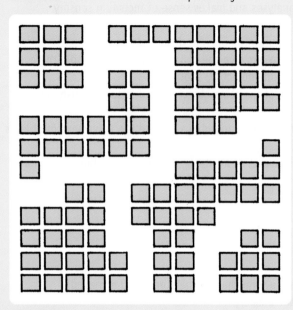

Figure 18.1

2   The distance of one or more objects from the eye is indicated by various types of visual cue. Which of the following visual cues is ABSENT from Figure 18.2?

A   superimposition   B   relative size
C   texture gradient   D   relative height in horizontal field

Figure 18.2

3 Which row in Table 18.1 correctly identifies blanks 1 and 2 in the following sentence?

The two different images of the same object sent by the eyes to the brain become merged, enabling the viewer to perceive a ____1____ image that allows him/her to judge the ____2____ of the object.

	Blank 1	Blank 2
A	monocular	size
B	binocular	size
C	monocular	distance
D	binocular	distance

Table 18.1

Questions 4 and 5 refer to the following possible answers.

- A perceptual constancy
- B binocular disparity
- C stereoscopic vision
- D depth perception

4 When a pencil, held 300 mm away from the face, is viewed alternately by each eye, it seems to move from side to side. What name is given to this phenomenon?

5 Figure 18.3 shows a window being opened. The window appears as three different-shaped images but the brain perceives it as an unchanging object. What name is given to this phenomenon?

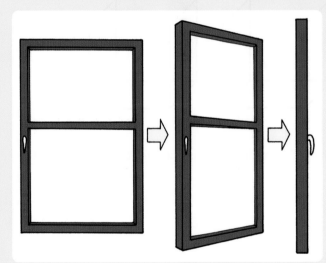

Figure 18.3

6 An experiment was set up to investigate the effect of previous experience on perceptual set. Before viewing the ambiguous image, group X were shown the cards marked set X and group Y were shown the cards marked set Y, as illustrated in Figure 18.4. Which part of Figure 18.5 was the ambiguous image used in this experiment?

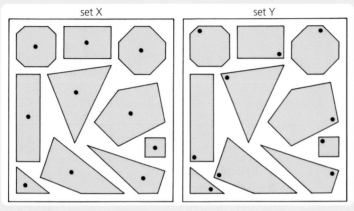

Figure 18.4

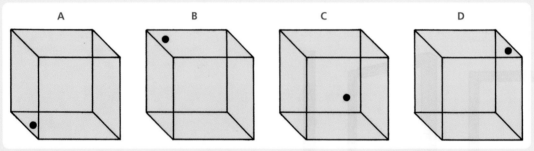

Figure 18.5

Questions 7, 8 and 9 refer to Figure 18.6, which shows a simplified version of the possible relationship between the short-term memory (STM) and long-term memory (LTM).

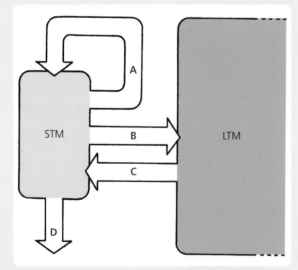

Figure 18.6

7 Which arrow represents an item being retrieved?

8 Which arrow represents an item being forgotten?

9 Which arrow represents an item being rehearsed?

10 Which of the following statements about short-term memory is NOT correct?
   A   It has a limited capacity of about seven items.
   B   Most of the images encoded in it are visual or auditory.
   C   Items are normally held for about 30 seconds only.
   D   Retrieval of items is inaccurate and unreliable.

11 Which of the following statements about long-term memory is correct?
   A   It stores various categories of semantic memories only.
   B   It holds items temporarily until they are retrieved by the STM.
   C   It has potential to retain an unlimited amount of information.
   D   It can retrieve items discarded by the STM.

Questions 12 and 13 refer to the following information. In an investigation into memory span, a group of 100 normal, healthy 17-year-old students were asked to listen to and then attempt to write down each of several series of letters. They started with a series of three letters and it then increased by one letter each time. The complete procedure was repeated twice using different lists of letters.

12 Which of the graphs in Figure 18.7 best represents the results of the investigation?

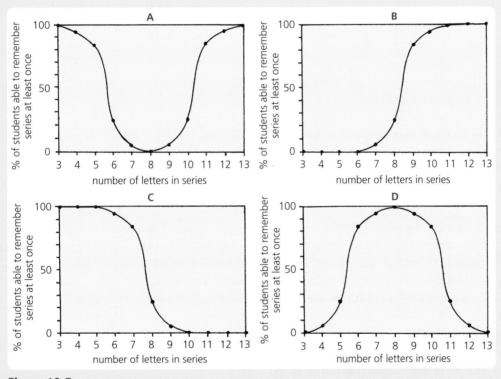

Figure 18.7

13 The most important reason for including 100 students in the investigation was to
   A   increase the reliability of the results.
   B   allow percentages to be easily calculated.
   C   obtain more accurate results for each subject's best score.
   D   ensure that age did not introduce a second variable factor.

14 Imagine that you meet for the first time in many years an old friend of your own age who used to be in your class in primary school and in the same sports team. If you tried to recall their name by checking your memory files on 'primary school' and 'sports team' you would be employing

A elaborate clues.

B displacement devices.

C contextual cues.

D organised chunks.

Questions 15 and 16 refer to the following information. In an investigation into the effect of self-recitation on memory, a large number of student volunteers were given a list of nonsense words to learn. The students were divided up into four groups and told how to use the study time of 15 minutes as shown in Table 18.2, which also gives the results of the experiment.

Group	Percentage of study time used to:		Percentage of nonsense words recalled:	
	silently read words to oneself	recite words to oneself	immediately	4 hours later
A	20	80	75	49
B	40	60	58	37
C	60	40	55	28
D	80	20	50	25

Table 18.2

15 Which group spent 6 minutes silently reading the words and 9 minutes reciting them?

16 From the investigation it can be concluded that

A reciting nonsense words helped at once to transfer them from LTM to STM.

B reading words silently to oneself was a more effective aid to memory than reciting them.

C self-recitation improved memory but the memorised words soon began to fade.

D spending more time silently reading the words results in a gradual improvement in recall after 4 hours.

17 When effective, self-recitation acts as a form of

A rehearsal.

B chunking.

C elaboration.

D organisation.

18 When trying to commit to memory the meaning of the words 'inflammatory response', you might simply choose 'defence mechanism' or you might explore the idea more fully and think of an inflammatory response as 'a localised defence mechanism in response to a minor injury where damaged mast cells beneath the skin surface release a chemical that causes vasodilation and is accompanied by a series of events that attract extra white blood cells to the site of infection'. If you chose the second option you would be aiding your memory by employing

A chunking.

B rehearsal.

C organisation.

D elaboration.

19 When memory is functioning properly

A encoding leads to storage of information that can be retrieved later.

B encoding leads to retrieval of information that can be stored later.

C retrieval leads to storage of information that can be encoded later.

D retrieval leads to encoding of information that can be stored later.

20 When attempting to memorise 20 similar objects viewed one after the other, people normally have most difficulty remembering the objects shown

A at the start of the display.

B in the middle of the display.

C towards the end of the display.

D both at the start and the end of the display.

21 The type of memory pattern referred to in question 20 is called the

A recent receptor effect.

B primary recall effect.

C immediate object effect.

D serial position effect.

22 Which of the following is NOT recommended by psychologists as an aid to learning an important piece of school work?

   A Paying very close attention to all of the text and diagrams.

   B Concentrating the learning process into one long session with a break at the end.

   C Rehearsing the key parts of the material until they are word-perfect.

   D Dividing learning time up into short sessions with breaks for rest.

23 An example of an episodic memory is remembering

   A how to dance the flamenco.

   B the phone number of your best friend.

   C how to skim-read a long passage of information.

   D the skills required to operate a word processor.

24 An example of a procedural memory is remembering the

   A movements that you need to make while swimming the crawl.

   B appearance of your suitcase at the airport luggage reclaim.

   C first names of all your aunts, uncles and cousins.

   D colour scheme of your bedroom before it was last redecorated.

25 Table 18.3 refers to possible locations of different types of memory in the brain. Which row is NOT correct?

	Type of memory	Possible location in the brain
A	emotional	limbic system and cerebral cortex
B	spatial	limbic system
C	procedural	motor region of cerebral cortex
D	semantic	medulla region of cerebellum

Table 18.3

19 The cells of the nervous system and neurotransmitters at synapses

Matching Test Part 1
Match the terms in list X with their descriptions in list Y.

list X
1 acetylcholine
2 axon
3 cell body
4 dendrite
5 glial
6 inter neuron
7 motor neuron
8 myelin sheath
9 myelination
10 neurotransmitter
11 node
12 norepinephrine
13 receptor site
14 sensory neuron
15 synapse
16 vesicle

list Y
a) nerve cell that carries nerve impulses from receptors in sensory organs to the CNS
b) nerve cell that passes nerve impulses from a sensory to a motor neuron
c) nerve fibre that carries nerve impulses towards a cell body
d) layer of fatty material forming an insulating jacket around axons
e) nerve fibre that carries nerve impulses away from a cell body
f) tiny region of functional contact between two neurons briefly connected by a neurotransmitter
g) part of a neuron that contains the nucleus and most of the cytoplasm
h) nerve cell that carries nerve impulses from the CNS to muscles or glands
i) neurotransmitter that undergoes degradation and re-uptake after transmission of an impulse at a synapse
j) development of myelin round axon fibres of neurons
k) type of cell that supports and maintains a homeostatic environment around neurons
l) small gap in a myelin sheath along an axon
m) neurotransmitter that is reabsorbed directly after transmission of an impulse at a synapse
n) location on the membrane of a postsynaptic neuron containing molecules with which a neurotransmitter briefly combines
o) structure in a presynaptic terminal containing stored neurotransmitter
p) general name for a chemical that allows transmission of a nerve impulse across a synapse

Matching Test Part 2
Match the terms in list X with their descriptions in list Y.

list X
1 agonist
2 antagonist
3 converging
4 desensitisation
5 diverging
6 dopamine
7 endorphins
8 plasticity
9 reverberating
10 sensitisation
11 summation
12 tolerance

list Y
a) cumulative effect of a series of weak stimuli that together bring about a nerve impulse
b) type of neural pathway where the route along which a nerve impulse passes is found to divide, transmitting it to several destinations
c) arrangement where neurons later in a pathway have axon branches that form synapses with neurons earlier in the pathway, enabling impulses to be recycled
d) type of neural pathway in which impulses from several sources are channelled towards a common destination
e) ability of brain cells to become altered as a result of new environmental experiences
f) neurotransmitter-like chemicals that act as natural painkillers
g) neurotransmitter that induces feelings of pleasure and reinforces beneficial behaviour
h) chemical that mimics the action of a neurotransmitter by blocking receptors on postsynaptic membranes
i) chemical that prevents the action of a neurotransmitter by binding to and stimulating receptors on postsynaptic membranes
j) state developed by a drug user whose reaction to the drug has decreased while the concentration of the drug has remained unchanged
k) process involving an increase in number and sensitivity of receptors as a result of repeated exposure to a drug acting as an antagonist
l) process involving a decrease in number and sensitivity of receptors as a result of repeated exposure to a drug acting as an agonist

Multiple Choice Test

Choose the ONE correct answer to each of the following multiple choice questions.

Questions 1–5 refer to Figure 19.1.

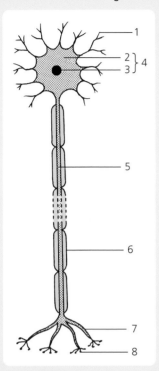

Figure 19.1

1 Which type of neuron is shown in Figure 19.1?
   A inter   B motor   C sensory   D association

2 Which numbered structure is the cell body?
   A 2   B 3   C 4   D 8

3 Which structure receives nerve impulses and passes them towards the cell body?
   A 1   B 5   C 7   D 8

4 The structure that contains vesicles of neurotransmitter is numbered
   A 1   B 2   C 3   D 8

5 The presence of which of the following structures greatly increases the speed at which the nerve impulses can be transmitted along an axon?
   A 1   B 3   C 6   D 7

6 Four stages (P, Q, R and S) that occur during the myelination of an axon by a glial cell are shown in Figure 19.2. What is their correct sequence?

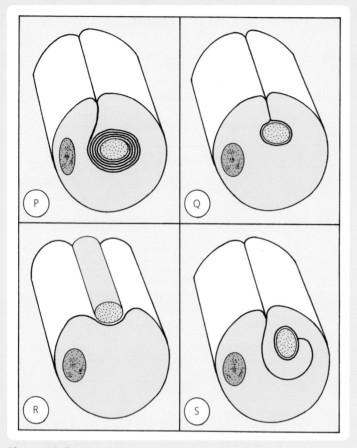

Figure 19.2

A Q, R, P, S   B Q, R, S, P   C R, Q, P, S   D R, Q, S, P

7 Which of the following statements about glial cells is NOT correct?
  A They provide neurons with physical support and some essential chemicals.
  B They transmit nerve impulses in association areas of the cerebral cortex.
  C They maintain a homeostatic environment around neighbouring neurons.
  D They remove debris and foreign material from the CNS by phagocytosis.

8 The direction in which a nerve impulse travels is always
  A dendrites ⟶ cell body ⟶ axon.
  B axon ⟶ cell body ⟶ dendrites.
  C cell body ⟶ dendrites ⟶ axon.
  D dendrites ⟶ axon ⟶ cell body.

Questions 9, 10 and 11 refer to Figure 19.3, which shows a synapse.

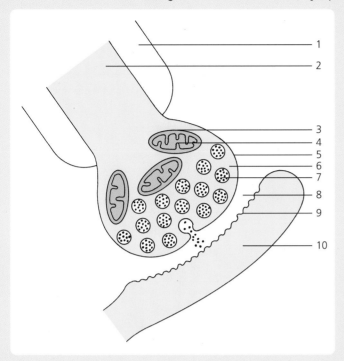

Figure 19.3

9 The synaptic cleft is indicated by number
   A 1   **B** 2   **C** 8   **D** 10

10 Which number points to a vesicle containing neurotransmitter?
   A 3   **B** 7   **C** 8   **D** 10

11 Which numbered label indicates the location of the receptors?
   A 4   **B** 5   **C** 6   **D** 9

12 Which of the following correctly represents the fate of a neurotransmitter after the transmission of a nerve impulse at a synapse?

   A acetylcholine $\xrightarrow{\text{enzyme}}$ non-active products   **B** acetylcholine $\xrightarrow{\text{enzyme}}$ noradrenaline

   C norepinephrine $\xrightarrow{\text{enzyme}}$ acetylcholine   **D** norepinephrine $\xrightarrow{\text{enzyme}}$ non-active products

13 It is NOT correct to say that a synapse is
   A a cytoplasmic connection between two types of nerve cell in a neural circuit.
   B located between an axon ending of one neuron and a dendrite of another.
   C only able to allow the transmission of nerve impulses in one direction.
   D a region of functional contact that can be bridged chemically by a neurotransmitter.

14 The immediate effect of applying an enzyme inhibitor that affects cholinesterase would be
   A increased release of acetylcholine from vesicles in synaptic terminals.
   B decreased release of acetylcholine from vesicles in synaptic terminals.
   C increased build-up of acetylcholine in synaptic clefts.
   D decreased build-up of acetylcholine in synaptic clefts.

Questions 15 and 16 refer to the possible answers shown in Figure 19.4.

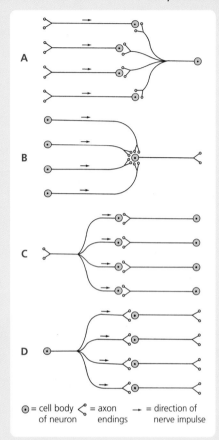

○ = cell body of neuron   ⟨ = axon endings   → = direction of nerve impulse

Figure 19.4

15 Which diagram represents a diverging neural pathway?
16 Which diagram represents a converging neural pathway?

17 In the reverberating neural pathway shown in Figure 19.5, an impulse could be sent back through the circuit via all of the following EXCEPT

A 3 ———→ 1

B 3 ———→ 2

C 4 ———→ 1

D 4 ———→ 3

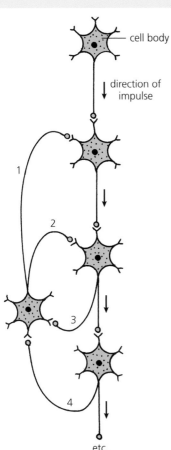

Figure 19.5

18 In an investigation, one group of rats (X) was reared in an unenriched environment and another group (Y) in an enriched environment. The rats were, in turn, released individually in a maze several times a day for 3 weeks. The bar graph in Figure 19.6 shows the results for the final maze run.

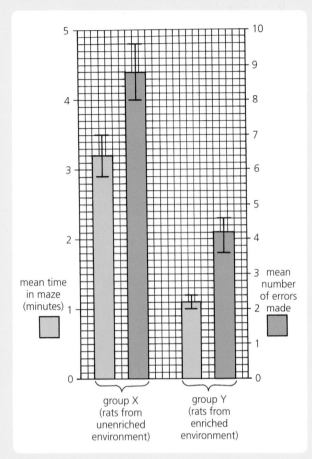

Figure 19.6

Which row in Table 19.1 gives two results that can be stated with a 95% level of confidence?

	Minimum mean time spent in maze by rats from unenriched environment (minutes)	Maximum mean number of errors made by rats from enriched environment
A	2.9	4.3
B	2.9	4.6
C	3.2	4.3
D	3.2	4.6

Table 19.1

19 Minor plasticity of response, for example successfully suppressing the blinking reflex, is thought to occur as a result of conflicting nerve messages meeting in a
   A   convergent pathway and the overall effect at the synapses being inhibitory.
   B   convergent pathway and the overall effect at the synapses being excitatory.
   C   divergent pathway and the overall effect at the synapses being inhibitory.
   D   divergent pathway and the overall effect at the synapses being excitatory.

20 Figure 19.7 shows a simplified version of a tiny part of the retina of the human eye. Which nerve fibre will transmit an impulse to the brain when light of VERY low intensity falls on the retina?

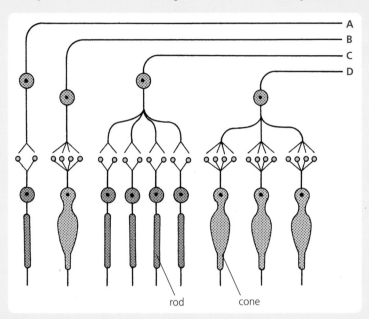

rod   cone

Figure 19.7

21 Which of the following statements is NOT correct?

Depending upon a person's circumstances, increased levels of endorphins may bring about

A regulation of appetite.   B feelings of euphoria.

C release of sex hormones.   D craving for chocolate.

22 Which row in Table 19.2 correctly identifies blanks 1 and 2 in the following sentence?

The reward pathway in the _____1_____ involves neurons that secrete the neurotransmitter _____2_____ following behaviour that is beneficial to the person.

	Blank 1	Blank 2
A	cerebral cortex	endorphin
B	limbic system	dopamine
C	cerebral cortex	dopamine
D	limbic system	endorphin

Table 19.2

23 Figure 19.8 shows the action of a normal natural neurotransmitter in the absence of drugs. Which part of Figure 19.9 correctly shows an antagonist drug in action?

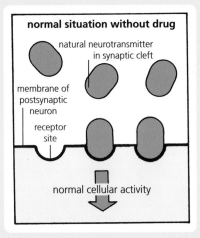

Figure 19.8

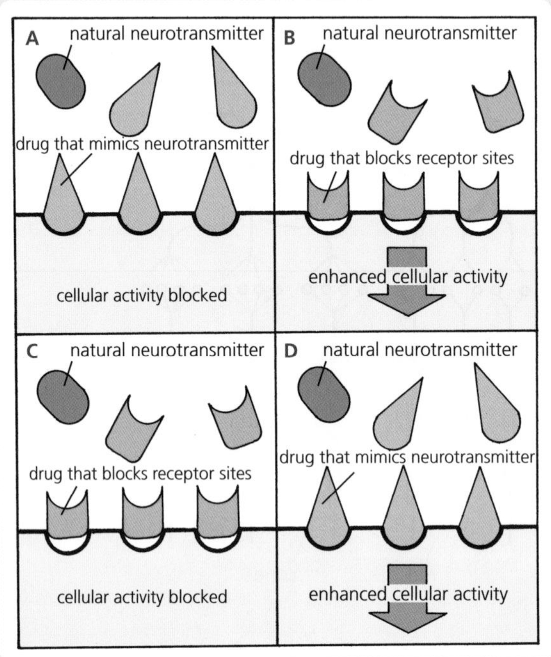

Figure 19.9

24 Figure 19.10 shows the results of monitoring the blood alcohol level of a man who drank 1.5 pints of beer at lunchtime and 2 pints after work on the same day.

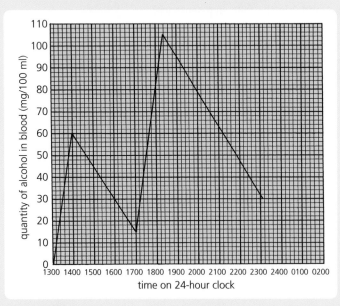

Figure 19.10

What percentage increase in alcohol content in his blood resulted from the after-work drinks?

A 60   B 90   C 600   D 900

25 When a drug user's reaction to an addictive drug is found to have decreased in intensity in response to the usual dose, the person is said to have built up

A a tolerance.   B an immunity.

C a resistance.   D a sensitisation.

20 Communication and social behaviour

Matching Test Part 1

Match the terms in list X with their descriptions in list Y.

list X
1. authoritarian
2. authoritative
3. body language
4. cognitive
5. communication
6. eye contact
7. facial expression
8. infant attachment
9. permissive
10. personal space
11. socialisation
12. socially competent
13. verbal language

list Y
a) strong emotional tie felt by an infant towards their primary carers
b) exchange of information, ideas and feelings between humans by verbal and non-verbal means
c) system that combines basic sounds into spoken words intelligible to others
d) non-verbal form of communication using the face to convey emotion or attitude
e) maintenance of gaze between two people
f) invisible territorial 'bubble' around a person that is respected by others
g) non-verbal form of communication using body posture or gestures to convey emotion or attitude
h) describing a person who possesses attributes that enable successful social interaction with others
i) gradual modification of a developing individual's behaviour to meet the social demands of their community
j) term used to refer to the ability to acquire knowledge and skills using intuition and reasoning
k) demanding but responsive form of parental control
l) excessively lenient form of parental control
m) unreasonably strict form of parental control

➡

Multiple Choice Test Part 1

Choose the ONE correct answer to each of the following multiple choice questions.

Questions 1 and 2 refer to Figure 20.1, which relates to social attachment in infants.

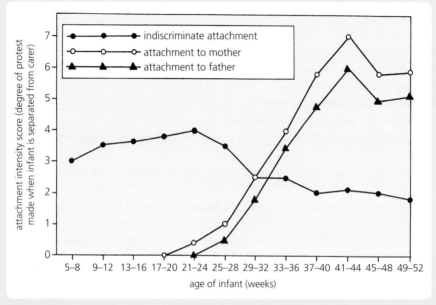

Figure 20.1

1 At what age in weeks did specific attachment to the father begin?

A 5–8   B 17–20   C 21–24   D 45–48

2 Which row in Table 20.1 correctly relates the age of the infant to the type of attachment to the carer?

	Age of infant (months)	Attachment to carer	
		type	present or absent
A	0–4	indiscriminate	absent
B	0–4	specific	present
C	5–8	indiscriminate	absent
D	5–8	specific	present

Table 20.1

Questions 3, 4, 5 and 6 refer to the following information. In each of a series of experiments, infant monkeys were given the choice between two surrogate mothers as shown in Figure 20.2. The number at the foot of each 'mother' indicates the average time (in hours per day) spent by an infant on the 'mother'.

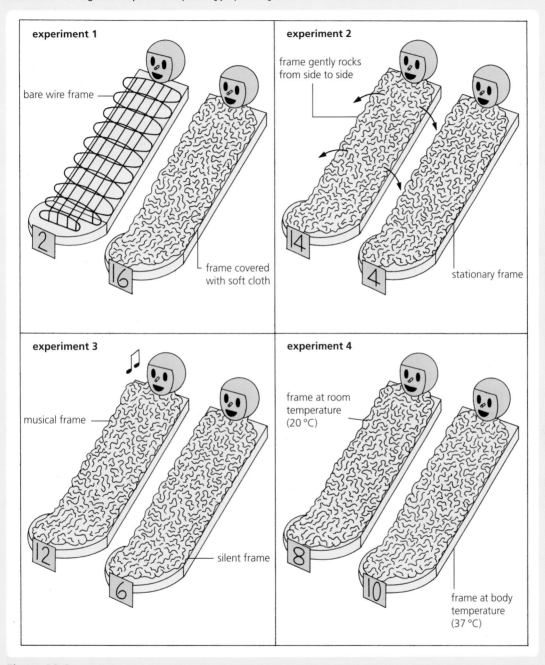

Figure 20.2

3 How many hours per day did an infant monkey spend exploring the surroundings away from the two surrogate mothers?

A 2   B 6   C 10   D 18

4 Which experiment demonstrates the need of an infant monkey for contact comfort?

A 1   **B** 2   **C** 3   **D** 4

5 From the given data, which variable was LEAST convincing as a factor that significantly played a part in the choice of mother by an infant monkey?

A cloth covering
C sound

B movement
D temperature

6 Which of the following combinations affecting the mother would be most likely to be favoured by infant monkeys?

A cloth, silent, moving, room temperature
C musical, body temperature, cloth, stationary

B body temperature, stationary, musical, bare wire
D stationary, cloth, silent, body temperature

Questions 7, 8 and 9 refer to the 'strange situation', a research tool devised to investigate infant attachment.

7 Human infants respond to the 'strange situation' in different ways depending on the type of attachment that exists between infant and mother. The type of attachment demonstrated by children who display major distress in response to the departure of the mother is found to be either

A secure or detached.
C avoidant or resistant.

B detached or avoidant.
D resistant or secure.

8 When the mother returns after a brief absence, a securely attached infant

A both seeks and resists comfort at the same time.
B approaches the mother, all the time looking away from her.
C goes straight to the mother and is quickly comforted.
D ignores the mother and continues to play with available toys.

9 Psychologists are of the opinion that an insecurely attached infant

A always accepts comfort from a stranger.
B will often suffer long-lasting effects.
C will become a warm and affectionate adult.
D always resists comfort from a stranger.

10 The more secure the attachment of an infant to their mother, the more likely that the child will

A investigate their immediate environment and develop their cognitive abilities.
B ignore their immediate environment and mimic their mother's body language.
C investigate their immediate environment and mimic their mother's body language.
D ignore their immediate environment and develop their cognitive abilities.

11 Which of the following methods of control used by parents is thought to result in the greatest level of social competence shown by their children?

A permissive
C indulgent

B authoritarian
D authoritative

Questions 12 and 13 refer to Figure 20.3.

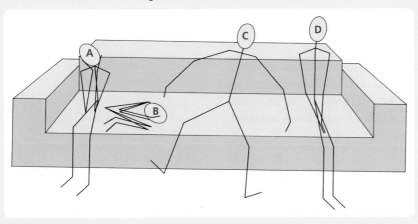

Figure 20.3

12 Which person's posture suggests a relaxed attitude?

13 Which person's posture suggests that he is terrified?

Questions 14 and 15 refer to Table 20.2.

	Facial expression	Body language	Tone of voice
A	✓	✓	✓
B	✓		✓
C		✓	✓
D			✓

Table 20.2

14 Which row in the table indicates the means of communication possible using a traditional land-line telephone?

15 Which row shows the means of communication possible using a webcam?

Questions 16 and 17 refer to Figure 20.4, which shows the results from an investigation into the effect of physical proximity on eye contact between a female interviewer and male interviewees.

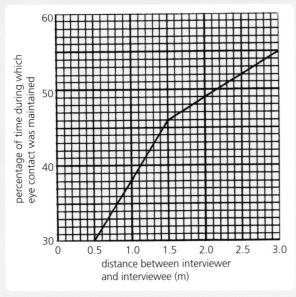

Figure 20.4

16 At distances of 1 metre and 2 metres between interviewer and interviewee, the respective percentage eye contact maintained was

   A 38 and 49   **B** 46 and 49   **C** 38 and 52   **D** 46 and 52

17 Which of the following conclusions is valid?

   A As distance decreases, percentage eye contact maintained increases.

   B As distance decreases, percentage eye contact maintained decreases.

   C As distance increases, percentage eye contact maintained decreases.

   D As distance increases, percentage eye contact maintained levels off.

18 A poor communicator would be one who

   A adopts a pleasant tone of voice.   **B** speaks as quickly as possible.

   C varies the tone of their voice.   **D** speaks clearly and concisely.

19 Which row in Table 20.3 best matches the auditory signal in each speaker's mode of delivery with their likely frame of mind?

	Auditory signal in speaker's mode of delivery		
	speaker 1 – loud voice	speaker 2 – high-speed delivery	speaker 3 – monotonous voice
A	angry	nervous	excited
B	tired	calm	excited
C	angry	nervous	bored
D	tired	calm	bored

Table 20.3

20 Which row in Table 20.4 correctly identifies blanks 1 and 2 in the following sentence?

Written language promotes the acceleration of _____1_____ and the development of culture and _____2_____ evolution.

	Blank 1	Blank 2
A	learning	social
B	infant attachment	social
C	learning	natural
D	infant attachment	natural

Table 20.4

Matching Test Part 2
Match the terms in list X with their descriptions in list Y.

list X
1 deindividuation
2 discrimination
3 extinction
4 generalisation
5 group pressure
6 identification
7 imitation
8 internalisation
9 motivation
10 practice
11 reinforcement
12 reward
13 shaping
14 social facilitation

list Y
a) process that makes an organism tend to repeat a certain piece of behaviour
b) process by which a skill or attitude is learned by copying a role model
c) procedure by which successive approximations of a desired behaviour pattern are reinforced
d) disappearance of a learned behaviour pattern caused by the withdrawal of its reinforcement
e) process by which repeated attempts improve the rate of learning of a motor skill
f) ability to distinguish between different but related stimuli and give different responses
g) inner drive that makes an animal want to participate in the learning process
h) ability to respond in the same way to many different but related stimuli
i) production of an increased performance in a competitive situation that involves carrying out a familiar activity
j) process by which an individual incorporates into herself a set of beliefs or values
k) an individual's loss of personal identity in a group that leads to diminished restraints on his behaviour
l) process by which an individual deliberately changes her beliefs to be like another person whom she admires
m) influence exerted by several people over one person, making that person abandon his views and adopt those of the others
n) reinforcer that increases the probability of a behavioural response being repeated

Multiple Choice Test Part 2
Choose the ONE correct answer to each of the following multiple choice questions.

1 Which of the set-ups in Figure 20.5 would be most suitable for use as a finger maze to investigate the effect of experience on learning?

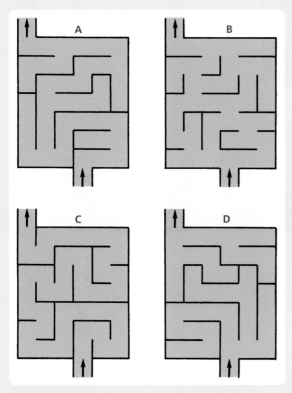

Figure 20.5

2 Table 20.5 lists some aspects of good practice carried out during a finger maze investigation. Which practice is correctly paired with the reason for carrying it out?

	Good practice	Reason
A	ten trials per learner	to ensure that no second variable factor is included in the investigation
B	experiment repeated with many learners	to obtain a more reliable set of results
C	learner blindfolded throughout all ten trials	to prevent two fingers being used at once
D	same learner used for each group of ten trials	to ensure that the recorded results are accurate

Table 20.5

3 Which of the graphs in Figure 20.6 correctly represents a learning curve?

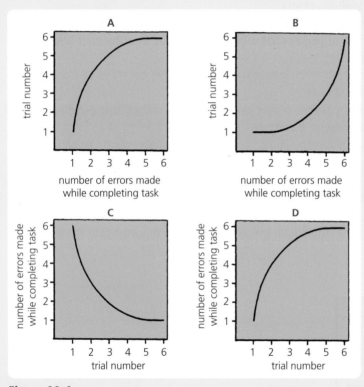

Figure 20.6

4 Which combination of aids shown in Table 20.6 is preferred by most people when learning a new motor skill?

	Written instructions to follow	Live expert to copy	Demonstration in the form of a long, single session	Demonstration made up of many short steps
A	✓		✓	
B		✓	✓	
C	✓			✓
D		✓		✓

Table 20.6

5 When given the choice in the Y-shaped maze shown in Figure 20.7, rats consistently choose the multi-alley side. It is possible that they are motivated by the desire to

A find a food supply.   B find a clean habitat.   C satisfy thirst.   D satisfy curiosity.

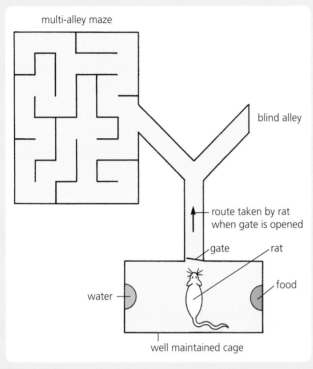

Figure 20.7

Questions 6, 7 and 8 refer to the graph in Figure 20.8. It represents the results from a learning experiment where four groups of rats (A, B, C and D) were tested in a maze (which contained several choices, each offering a correct and a wrong route).

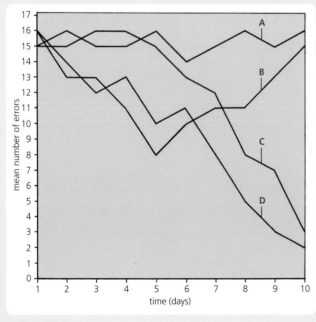

Figure 20.8

6 In which group were the rats NEVER motivated?

7 Which group of rats received food rewards from day five onwards?

8 Which group of rats lost motivation after day five?

9 The hungry chimpanzee shown in Figure 20.9 learns that by piling up the boxes in a certain way he can reach the banana. If he is allowed to eat the banana each time he reaches it, then it is correct to say that the

A   response is reinforced and the organism is rewarded.

B   response is reinforced and the organism is reinforced.

C   response is rewarded and the organism is reinforced.

D   response is rewarded and the organism is rewarded.

Figure 20.9

10 The girl in Figure 20.10 is learning to brush her teeth. This form of behaviour, which develops as a result of her parents reinforcing successive approximations of the desired response, is called

A   generalising.   B   discriminating.   C   imitating.   D   shaping.

Figure 20.10

11 A young woman was severely scratched by a Siamese cat when she was a child. Now she is scared of all cats. This is an example of

  A extinction.   **B** generalisation.   **C** discrimination.   **D** reinforcement.

12 Which of the following statements is NOT correct?

A social group consists of people who normally

  A behave according to an agreed set of rules.

  B communicate and interact with one another over an extended period of time.

  C act independently so that an event affecting one rarely affects the others.

  D share certain interests and goals that can be achieved by working together.

13 Table 20.7 shows four different situations involving a worker in a car assembly plant.

Under which set of conditions would the worker be most likely to produce his best performance of the task set?

	Task set		In competition with other workers?	
	familiar	unfamiliar	yes	no
A	✓		✓	
B	✓			✓
C		✓	✓	
D		✓		✓

Table 20.7

14 Which of the following is NOT a reward obtained from membership of a social group?

  A esteem   **B** praise   **C** attention   **D** privacy

Questions 15 and 16 refer to Figure 20.11, which shows the results of a group pressure experiment involving many volunteers in 11 different trials.

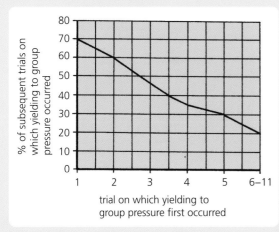

Figure 20.11

15 For a person who yielded to group pressure on trial 1, the NUMBER of subsequent trials on which he also yielded was

  A 1   **B** 7   **C** 7.7   **D** 70

16 If a person did not yield to group pressure until trial 5, what percentage of subsequent trials did she also yield on?

  A 5   **B** 6–11   **C** 30   **D** 50   ➡

17 Which of the following factors would a TV director try to AVOID in a programme aimed at persuading viewers to vote for a particular political party?

  A programme content that is interesting and occasionally humorous

  B presenter who repeatedly displays unusual mannerisms

  C programme content that presents both sides of the argument

  D presenter who has an attractive appearance and good vocal delivery

18 Destruction of a bus shelter by a gang of vandals shows a weakening of inner controls such as guilt and fear of punishment that normally prevent an individual from behaving in such a manner in isolation. Such group behaviour involving anonymity and diffusion of responsibility is called

  A identification.   **B** group pressure.

  C deindividuation.   **D** social facilitation.

19 Some young people are tempted to experiment with addictive drugs in the belief that it will be worth the risk involved. Drug education programmes often use posters to illustrate the downside of taking drugs. Which of the following terms is used to refer to such an attempt to alter a person's beliefs by persuasion?

  A group pressure   **B** identification

  C deindividuation   **D** internalisation

20 The young athlete in Figure 20.12 has stopped smoking and now spends several hours every week training to try to be like his hero. This form of behaviour is called

  A identification.   **B** deindividuation.

  C internalisation.   **D** social facilitation.

Figure 20.12

Unit 4

Immunology and Public Health

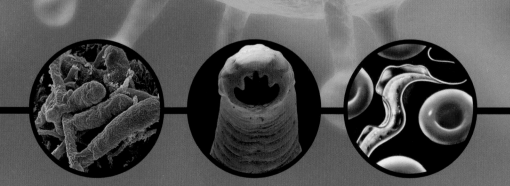

21 Non-specific defences

Matching Test

Match the terms in list X with their descriptions in list Y.

list X
1 antimicrobial proteins
2 apoptosis
3 chemical
4 clotting elements
5 cytokines
6 epithelial
7 histamine
8 immunity
9 inflammatory response
10 lysosome
11 mast
12 natural killer
13 non-specific
14 phagocyte
15 physical
16 vasodilation

list Y
a) ability of the body to resist infection by a pathogen
b) form of defence that acts against any type of invading pathogen
c) type of cell that offers physical protection against micro-organisms
d) type of defence of the body provided by epithelial cells and mucous membranes
e) type of defence of the body provided by lysozyme in saliva and acid in the stomach
f) localised defence mechanism at a site of injury that involves vasodilation and enhanced phagocytic activity
g) process by which the bore of a blood vessel becomes wider
h) type of cell closely related to blood cells that releases histamine
i) chemical released by injured mast cells that causes blood vessels to dilate during the inflammatory response
j) cell-signalling protein molecules secreted by many cell types, including white blood cells
k) chemicals that cause blood to coagulate at a site of injury
l) chemicals that amplify the immune response
m) process of programmed cell death following production by cell of self-destructive enzymes
n) non-phagocytic white blood cell that plays a non-specific defence role by causing the apoptosis of virus-infected cells
o) white blood cell that plays a non-specific defence role by engulfing and destroying pathogens
p) organelle present in a phagocyte that contains digestive enzymes such as lysozyme

Multiple Choice Test

Choose the ONE correct answer to each of the following multiple choice questions.

1 Which of the following statements are BOTH correct?
    1 Specific defences work against any pathogen.
    2 Each specific defence works against one particular type of pathogen only.
    3 Non-specific defences work against any pathogen.
    4 Each non-specific defence works against one particular type of pathogen only.
      A 1 and 3   **B** 1 and 4   **C** 2 and 3   **D** 2 and 4

2 The labels in Figure 21.1 indicate some of the structures that defend the body against pathogens.

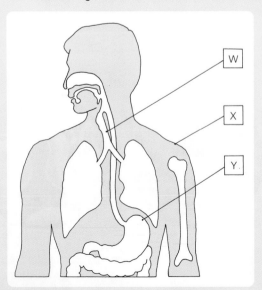

Which of the following provide the body with forms of non-specific defence?
A W and X only
B W and Y only
C X and Y only
D W, X and Y

Figure 21.1

3 All of the following are examples of non-specific defences EXCEPT
   A activity of phagocytic blood cells.   B secretion of acid by gastric glands.
   C formation of tears containing lysozyme.   D production of antibodies by lymphocytes.

Questions 4, 5 and 6 refer to Figure 21.2, which shows some of the events involved in the inflammatory response.

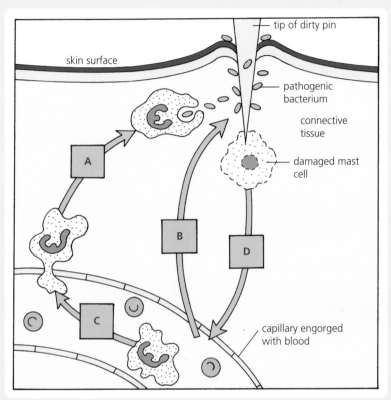

Figure 21.2

4 Which box represents 'increased flow of fluid from bloodstream to infected site'?
5 Which box represents 'release of histamine'?
6 Which box represents 'migration of phagocyte through connective tissue to infected area'?
7 During the inflammatory response, increased permeability of capillary walls promotes rapid
   A delivery of blood-clotting elements to the site of injury.
   B removal of antimicrobial proteins from the infected site.
   C migration of red blood cells to the infected wound.
   D vasodilation of neighbouring venules and capillaries.

8 Which of the following statements is NOT correct?
   Cytokines are
   A cell-signalling protein molecules.
   B also known as antibiotic histamines.
   C secreted by phagocytes and natural killer cells.
   D involved in specific and non-specific defences.

➡

9 Figure 21.3 shows four of the stages that occur during the process of phagocytosis.

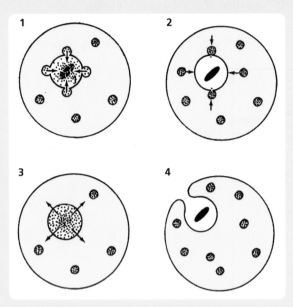

Figure 21.3

Which column in Table 21.1 correctly matches each of these numbered stages with its description?

Description of stage	A	B	C	D
products of digestion pass into the cytoplasm of the phagocyte	2	3	2	3
some lysosomes move towards and fuse with the vacuole	3	2	3	2
the phagocyte forms a vacuole around the bacterium	1	1	4	4
digestive enzymes break down the bacterium	4	4	1	1

Table 21.1

10 The boxed statements in Figure 21.4 give the steps involved in the apoptosis of a virus-infected cell.

P 'suicide' genes in the target cell become switched on

Q protein molecules released by the natural killer cell form pores in the membrane of the virus-infected cell

R 'suicide' genes code for degradative enzymes that bring about self-destruction of the cell

S signal molecules from the natural killer cell enter the infected target cell

Figure 21.4

Which of the following gives the correct sequence of events?

A Q, S, P, R   B Q, S, R, P   C S, Q, P, R   D S, Q, R, P

22 Specific cellular defences

Matching Test
Match the terms in list X with their descriptions in list Y.

list X
1 allergy
2 antibody
3 antigen
4 autoimmune disease
5 autoimmunity
6 B cell
7 clonal selection
8 cytokines
9 lysis
10 memory cells
11 pathogen
12 Rhesus negative
13 Rhesus positive
14 specific
15 T cell
16 toxin

list Y
a) cell-signalling protein molecules secreted by many cell types including white blood cells
b) process by which a lymphocyte is activated by an antigen and responds by dividing into a population of identical lymphocytes
c) type of lymphocyte that may secrete cytokines that activate other blood cells or may cause an infected cell to destroy itself
d) type of lymphocyte that on being stimulated produces clones of antibody-forming cells and memory cells
e) disease-causing micro-organism
f) poison produced by a pathogen
g) inability of the body to tolerate the antigens that make up the self message on its cell surfaces
h) condition such as multiple sclerosis caused by T lymphocytes attacking the body's own cells
i) over-reaction by hypersensitive immune system's B lymphocytes to a harmless substance
j) term used to describe people who make anti-D antibodies
k) term used to describe people who have antigen D on their red blood cells
l) bursting of diseased cell following an attack by white blood cells
m) B and T lymphocytes formed during first exposure to an antigen that produce a secondary response in the future
n) complex molecule recognised by the body as non-self and foreign
o) Y-shaped protein molecule whose arms each bear a receptor site specific to an antigen
p) term for the form of cellular defence that acts against one particular type of pathogen

Multiple Choice Test
Choose the ONE correct answer to each of the following multiple choice questions.

1 Which of the following can act as antigens?
Molecules that are present on or in
A viruses only.
B bacteria and viruses only.
C cancer cells and bacterial toxins only.
D viruses, bacteria, bacterial toxins and cancer cells.
2 Each lymphocyte has on the surface of its cell membrane
A a single copy of one type of antigen receptor.
B several copies of one type of antigen receptor.
C a single copy of each of several types of antigen receptor.
D several copies of each of several types of antigen receptor.

3 Which row in Table 22.1 correctly identifies blanks 1 and 2 in the following sentence?

When a lymphocyte becomes attached to and activated by an antigen, it is said to have undergone _____1_____ and it responds by forming a _____2_____.

	Blank 1	Blank 2
A	selection	clonal population
B	selection	degradative enzyme
C	apoptosis	clonal population
D	apoptosis	degradative enzyme

Table 22.1

4 Which row in Table 22.2 correctly indicates the type(s) of ANTIGEN present on the surfaces of the red blood cells of people with each of the four different blood groups?

	Blood group of person			
	A	B	AB	O
1	antigen A	antigen B	antigens A and B	neither antigen type
2	antigen B	antigen A	antigens A and B	neither antigen type
3	antigen A	antigen B	neither antigen type	antigens A and B
4	antigen B	antigen A	neither antigen type	antigens A and B

Table 22.2

A 1   B 2   C 3   D 4

5 Which row in Table 22.3 correctly indicates the type(s) of ANTIBODY present in the plasma of people with each of the different blood groups?

	Blood group of person			
	A	B	AB	O
1	anti-B antibody	anti-A antibody	anti-A and anti-B antibodies	neither type
2	anti-A antibody	anti-B antibody	anti-A and anti-B antibodies	neither type
3	anti-B antibody	anti-A antibody	neither type	anti-A and anti-B antibodies
4	anti-A antibody	anti-B antibody	neither type	anti-A and anti-B antibodies

Table 22.3

A 1   B 2   C 3   D 4

6 A person with the blood group AB can safely receive blood from people with group(s)

A AB only.   B A and B only.   C A, B and AB only.   D A, B, AB and O.

7 There is a chance that a baby suffering from haemolytic disease of the newborn (HDNB) could be produced by parents who are

A ♂ Rh– and ♀ Rh–   B ♂ Rh– and ♀ Rh+
C ♂ Rh+ and ♀ Rh–   D ♂ Rh+ and ♀ Rh+

Questions 8 and 9 refer to the following possible answers.

   A allergy   B apoptosis   C autoimmunity   D antigenic lysis

8 When the body no longer tolerates self-antigens on its cell surfaces, it may launch an attack on its own cells. What name is given to this failure in regulation of the immune system?

9 Which answer refers to the series of events shown in Figure 22.1?

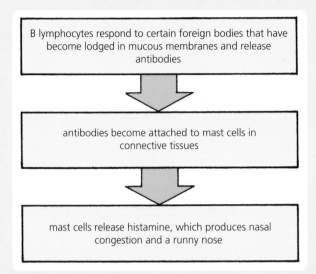

Figure 22.1

10 The graph in Figure 22.2 charts the estimated number of cases of occupational asthma for a European country.

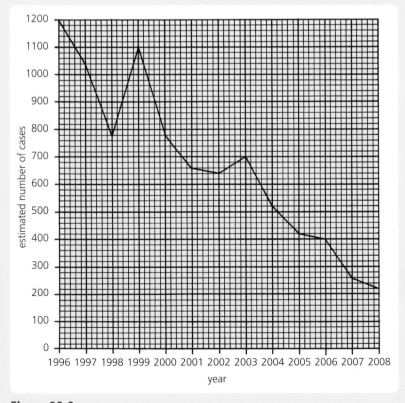

Figure 22.2

According to this estimate, the number of cases showed the greatest percentage decrease between years

   A 1996 and 1998.   B 1999 and 2001.   C 2003 and 2005.   D 2006 and 2008.

11 Which of the following statements does NOT apply to either group of T lymphocytes?
  A They destroy infected cells by inducing apoptosis.
  B They respond to the presence of antigens by producing antibodies.
  C They move to a site of infection under the direction of cytokines.
  D They secrete cytokines that activate B lymphocytes.

12 Figure 22.3 shows four stages (W, X, Y and Z) in the destruction of a cancer cell by a T lymphocyte.

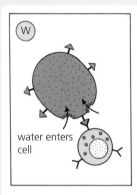
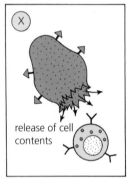
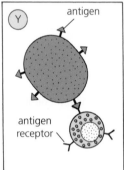
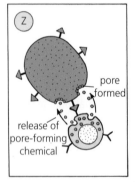

Figure 22.3

What is the correct order in which the stages occur?
  A Y ⟶ Z ⟶ X ⟶ W
  B Y ⟶ Z ⟶ W ⟶ X
  C Z ⟶ Y ⟶ X ⟶ W
  D Z ⟶ Y ⟶ W ⟶ X

13 Which of the four diagrams in Figure 22.4 shows the correct structure of an antibody molecule? (R = receptor site, N = non-receptor site)

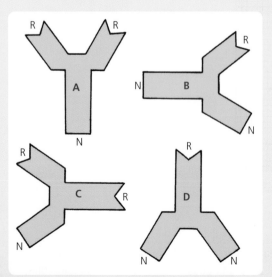

Figure 22.4

14 Table 22.4 gives information about five types of antibody found in the blood of a human population.

Antibody	I_gG	I_gA	I_gM	I_gD	I_gE
Mean serum concentration	12.0 g/l	2.8 g/l	1.3 g/l	0.2 g/l	0.7 mg/l

Table 22.4

The mean serum concentration of I_gA is greater than that of I_gE by a factor of

A 4 times.   B 40 times.   C 400 times.   D 4000 times.

15 When a virus with antigens on its surface invades an organism and multiplies, the events listed in Figure 22.5 occur.

1 B cell becomes activated and produces a clone of activated B cells

2 free antibodies combine with and contain the viral antigens

3 T cell recognises antigen on B cell and releases cytokines

4 activated B cells produce and release antibodies into bloodstream

5 B cell takes in and displays molecules of viral antigen

Figure 22.5

Their correct order is
A 3, 5, 1, 2, 4.   B 5, 3, 1, 4, 2.   C 5, 1, 3, 4, 2.   D 1, 3, 5, 2, 4.

16 Which of the following statements is NOT correct?
Antigen–antibody complexes may
A inactivate a toxin released by a pathogen.
B destroy a pathogen by the process of apoptosis.
C render a pathogen more susceptible to phagocytosis.
D stimulate a response that results in lysis of the pathogen.

17 Which of the graphs in Figure 22.6 best represents the primary and secondary responses on exposure to a pathogen?

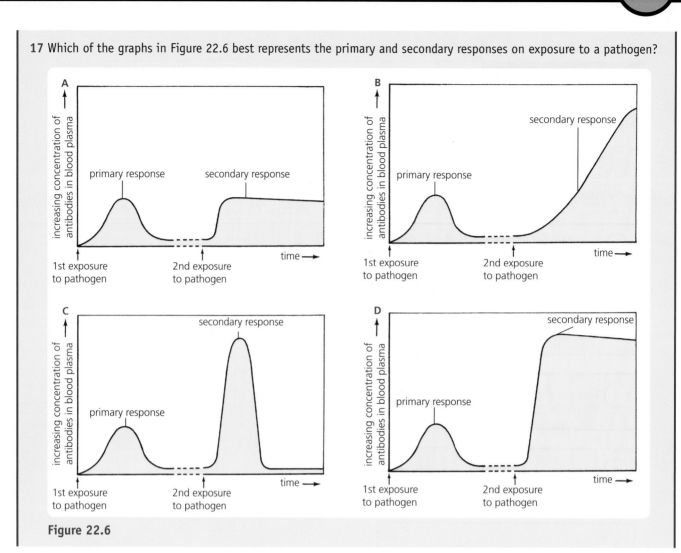

Figure 22.6

Questions 18, 19 and 20 refer to Figure 22.7, which summarises the specific immune response.

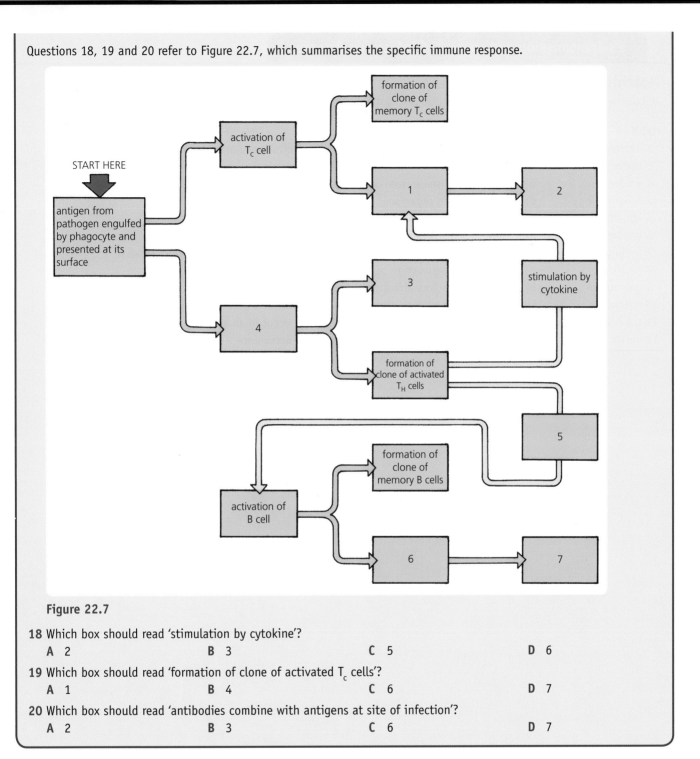

Figure 22.7

18 Which box should read 'stimulation by cytokine'?

A 2   B 3   C 5   D 6

19 Which box should read 'formation of clone of activated T_c cells'?

A 1   B 4   C 6   D 7

20 Which box should read 'antibodies combine with antigens at site of infection'?

A 2   B 3   C 6   D 7

23 Transmission and control of infectious diseases

Matching Test

Match the terms in list X with their descriptions in list Y.

list X

1 antisepsis
2 endemic
3 epidemic
4 epidemiology
5 hygiene
6 pandemic
7 pathogen
8 protozoa
9 quarantine
10 sporadic
11 transmission
12 vector

list Y

a) clean, healthy practices and habits that control populations of microbes present in and on the body
b) transfer of a disease from one person to another
c) disease-causing organism
d) single-celled animals
e) organism that carries a pathogen from one host to another
f) period of isolation of a person to prevent the spread of an infectious disease
g) inhibition or destruction of micro-organisms that cause disease
h) study of the factors affecting the spread of infectious diseases
i) disease spread pattern where a series of epidemics occurs throughout the world
j) disease spread pattern where regular cases occur in an area
k) disease spread pattern showing occasional occurrence
l) disease spread pattern where an unusually high number of cases occurs in an area

Multiple Choice Test

Choose the ONE correct answer to each of the following multiple choice questions.

1 Many years ago scientists set out to discover the means by which the bacteria that cause bubonic plague are spread among rats. They began with the hypothesis 'Fleas must be present for the bacteria to spread from rat to rat'. Then they set up the experiment shown in Figure 23.1.

Which set of results in Figure 23.2 supports their hypothesis?

rats fatally infected with plague and carrying fleas
rats not infected with plague and carrying fleas
placed together in cage X
cage X
rats fatally infected with plague but with all fleas removed
uninfected rats with all fleas removed
placed together in cage Y
cage Y

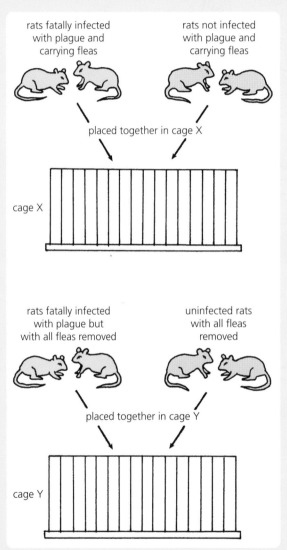

Figure 23.1

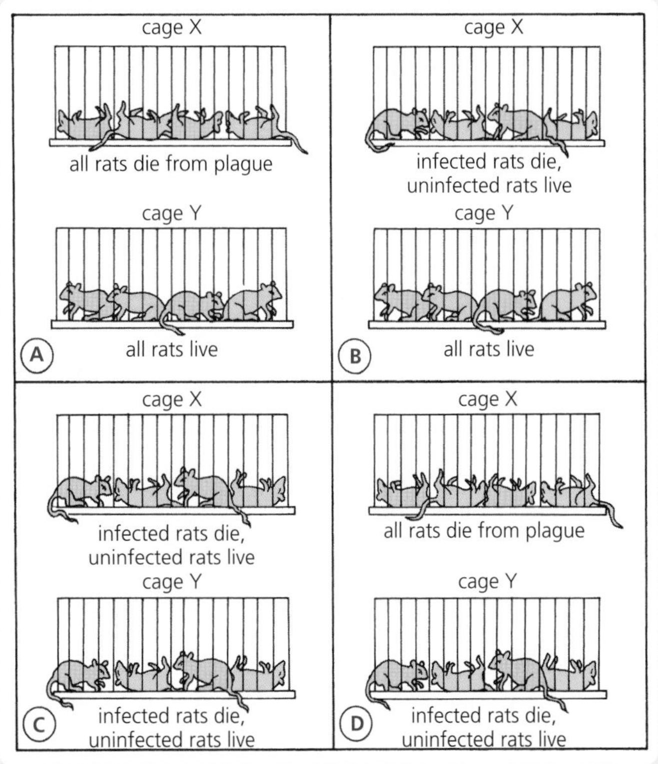

cage X
all rats die from plague
cage Y
all rats live
A
cage X
infected rats die, uninfected rats live
cage Y
all rats live
B
cage X
infected rats die, uninfected rats live
cage Y
infected rats die, uninfected rats live
C
cage X
all rats die from plague
cage Y
infected rats die, uninfected rats live
D

Figure 23.2

2   Figure 23.3 shows the life cycle of a liver fluke found in rural parts of Asia.

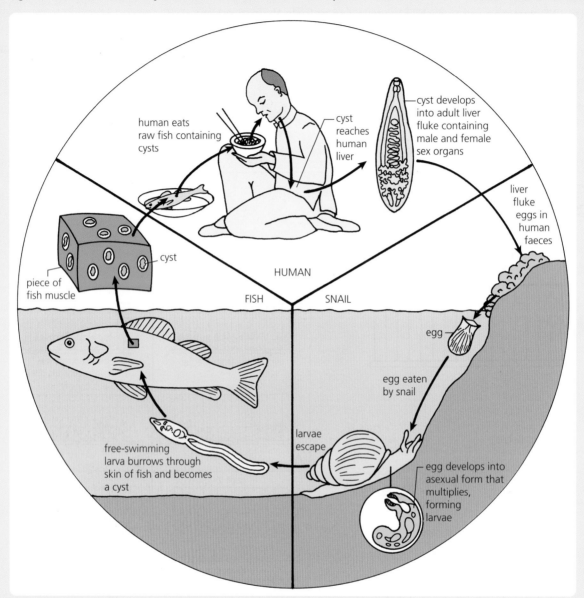

Figure 23.3

Which of the following procedures would be the easiest and most cost-effective method of breaking this cycle
of events?

A   removing all snails from the fish ponds using molluscicide

B   restocking the ponds with fish that eat snails

C   cooking all fish thoroughly before consumption by humans

D   building sewage treatment works beside the fish ponds

Questions 3 and 4 refer to the following information. An experiment was set up to investigate the effect of a digestive enzyme from the human host on the hatching of tapeworm cysts. The results are shown in Figure 23.4.

100
80
60
40
20
0
hatching of cysts (%)
intestinal fluid containing digestive enzyme
intestinal fluid lacking digestive enzyme
0
30
60
90
120
time (min)

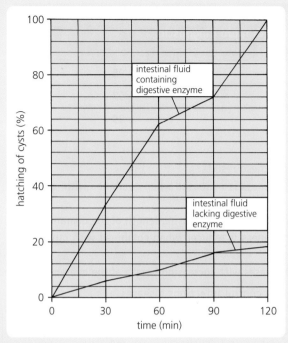

Figure 23.4

3 From the graph it can be concluded that at 1 hour 30 minutes, presence of the enzyme compared to lack of the enzyme brought about the hatching of more cysts by a factor of

A 3.5 times. **B** 4.5 times. **C** 5.5 times. **D** 56.0 times.

4 In this experiment the dependent variable factor was

A percentage hatching of cysts. **B** absence of digestive enzyme.
C time in minutes. **D** presence of digestive enzyme.

5 In 1890 the German scientist Robert Koch set out his famous criteria for deciding whether or not a given bacterium is the cause of a given disease. These so-called postulates (necessary conditions) are shown in Figure 23.5.

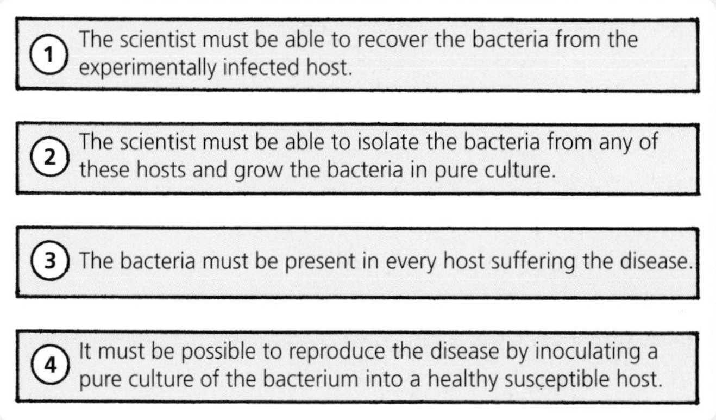

1 The scientist must be able to recover the bacteria from the experimentally infected host.
2 The scientist must be able to isolate the bacteria from any of these hosts and grow the bacteria in pure culture.
3 The bacteria must be present in every host suffering the disease.
4 It must be possible to reproduce the disease by inoculating a pure culture of the bacterium into a healthy susceptible host.

Figure 23.5

Their correct order is

A 2, 3, 1, 4 **B** 2, 3, 4, 1 **C** 3, 2, 1, 4 **D** 3, 2, 4, 1

6   The results in Table 23.1 refer to a survey on the quarantine procedures carried out at the borders of a small island nation to control the threat of an influenza pandemic.

Length of quarantine period (days)	Effectiveness (%)	Length of quarantine period when combined with rapid diagnostic testing (days)	Effectiveness (%)
4.6	95	2.6	95
6.6	97	4.1	97
8.6	99	5.7	99

Table 23.1

Which of the following conclusions can be correctly drawn from these results?

A   The length of the quarantine period needed for 95% effectiveness is increased when combined with rapid diagnostic testing.

B   The percentage effectiveness of the quarantine measures increases with decreasing length of the quarantine period.

C   The length of the quarantine period needed for 99% effectiveness is reduced when combined with rapid diagnostic testing.

D   The percentage effectiveness of the quarantine measures decreases with increasing length of the quarantine period.

Questions 7 and 8 refer to Table 23.2, which gives measures used to control infectious diseases.

A	inspection of food sources at farms	chlorination of drinking water
B	pasteurisation of milk	careful attention to sexual health
C	safe handling of food by each member of a household	regular washing of hands
D	good personal hygiene	effective large-scale system of waste disposal

Table 23.2

7   In which row of the table are BOTH methods of control the responsibility of the individual?
8   In which row of the table are BOTH methods of control the responsibility of the community?

Some methods of preventing spoilage of stored food by bacteria depend on the use of an appropriate temperature treatment. Questions 9 and 10 refer to Figure 23.6.

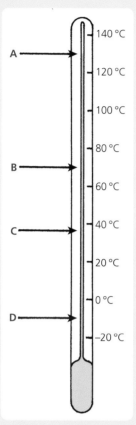

Figure 23.6

9 At which temperature would bacteria be alive but not growing?

10 At which temperature would many bacteria be killed but a few survive?

11 All of the following are methods used to try to control the number of mosquitoes carrying malaria in a region EXCEPT

  A draining stagnant water.   **B** introducing sterile males.

  C using larvicides and insecticides.   **D** draping netting over beds at night.

12 In the UK, systems of control are employed to make the entire process of food manufacture from 'stable to table' safe for the consumer. Which of the following is NOT a control measure?

  A humane killing practices   **B** traceability of food sources

  C scientific inspections   **D** risk analysis

13 Figure 23.7 presents data about the number of cases of influenza in a European city over a period of 4 years.

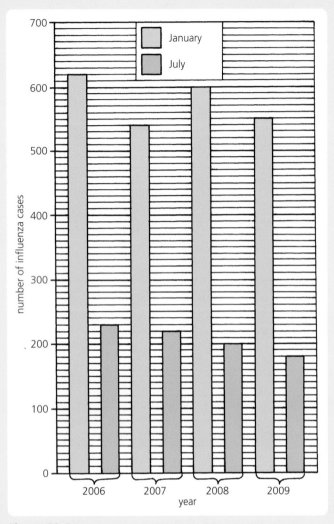

Figure 23.7

Between January and July of which year did the greatest PERCENTAGE decrease in number of influenza cases occur?

A 2006   B 2007   C 2008   D 2009

Questions 14, 15 and 16 refer to the following possible answers, each of which describes the spread pattern of an infectious disease.

A It occurs as series of epidemics that spreads across whole continents or even throughout the world.

B It occurs in scattered or isolated instances with no connection between them.

C It recurs as a regular number of cases in a particular area.

D It simultaneously affects an unusually large number of people in a particular area.

14 Which of these describes endemic spread of a disease?

15 Which of these describes pandemic spread of a disease?

16 Which of these describes sporadic spread of a disease?

17 An epidemiologist studies

   A epidemics but not pandemics.

   C spread patterns of infectious diseases.

   B sporadic effects of quarantine measures.

   D inherited disorders endemic in inbred populations.

18 Which of the following is LEAST likely as a measure that epidemiologists would suggest to control a disease epidemic?

   A introduce a programme of famine relief

   B prevent spread of pathogen by vector from person to person

   C immunise people not yet infected with vaccine

   D treat infected people with appropriate drugs

Questions 19 and 20 refer to Table 23.3. It contains data about a worldwide disease in 2008.

	Number suffering the disease at start of 2008	Number of new cases during 2008	Number of deaths from the disease during 2008	Total number suffering the disease at end of 2008
Europe	0.72×10^6	0.65×10^6	0.06×10^6	box X
World total	13.12×10^6	9.48×10^6	1.20×10^6	21.40×10^6

Table 23.3

19 The answer missing from box X is

   A 0.77×10^6 **B** 1.31×10^6 **C** 1.43×10^6 **D** 1.97×10^6

20 The number of deaths that occurred in Europe expressed as a percentage of the world total is

   A 2   **B** 5   **C** 20   **D** 50

24 Active immunisation and vaccination and the evasion of specific immune response by pathogens

Matching Test

Match the terms in list X with their descriptions in list Y.

list X
1 active immunisation
2 adjuvant
3 antigenic variation
4 bias
5 contact parameter
6 double-blind
7 efficacy
8 experimental error
9 herd immunity
10 herd immunity threshold
11 immunological memory
12 placebo
13 placebo effect
14 protocol
15 randomisation
16 statistically significant
17 vaccination
18 virulence

list Y
a) procedural method with a design and implementation that meet agreed standards
b) procedure used to eliminate bias
c) trial in which neither subjects nor doctors know who is receiving the drug being tested and who is receiving the placebo
d) change resulting from receiving the treatment that does not depend on the treatment's active ingredient
e) 'sham' treatment
f) irrational preference or prejudice
g) unintended, uncontrolled variability that exists between treatments in a trial
h) describing results that are reliable when subjected to statistical analysis and are not the result of chance
i) effectiveness of a vaccine or medicine
j) protection of non-immune minority of a population from a disease by the presence of the immune majority
k) percentage of immune individuals in a population above which a disease no longer manages to persist
l) capacity of a pathogen to cause disease
m) degree of host population density that affects a pathogen's ability to spread
n) state demonstrated by new strains of a pathogen with surface antigens different from those on the original strain
o) process by which a person develops immunity to a pathogen as a result of their body producing antibodies
p) deliberate introduction of a harmless form of a pathogen into the body to initiate the immune response
q) ability of the body's memory T and B cells to respond to reinfection by a pathogen and produce antibodies against it
r) substance that promotes the activity of the antigen in a vaccine and enhances the immune response

Multiple Choice Test

Choose the ONE correct answer to each of the following multiple choice questions.

1 Figure 24.1 shows a disease-causing virus.

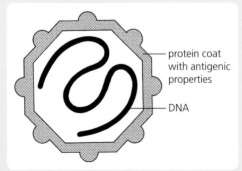

protein coat with antigenic properties

DNA

Figure 24.1

Which of the following could be safely used as a vaccine to initiate the immune response?

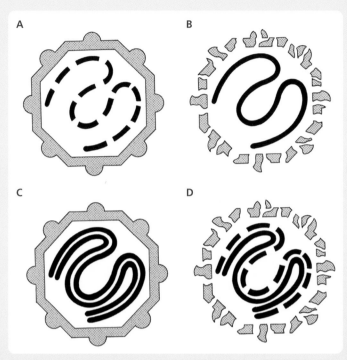

Figure 24.2

2 The form of immunity resulting from vaccination using an altered form of the pathogen is said to be

A   actively acquired by natural means.
C   passively acquired by natural means.

B   actively acquired by artificial means.
D   passively acquired by artificial means.

3 Figure 24.3 shows the stages involved in a clinical trial.

> (1) treatment tested on a large number of people (e.g. 300) who have the illness to check that it is safe and effective

> (2) licence sought from government to manufacture the treatment

> (3) treatment tested on a very large number of people (e.g. 2000) who have the illness, using standard protocols

> (4) post-market studies carried out to check for rare side effects and long-term risks

> (5) treatment tested on a small number of healthy volunteers (e.g. 50) to check that it is safe

Figure 24.3

The correct sequence is

A   3, 1, 5, 2, 4   B   5, 1, 3, 2, 4   C   3, 1, 5, 4, 2   D   5, 1, 3, 4, 2

Questions 4, 5 and 6 refer to the following possible answers, which are the reasons for including certain features when designing the phase III stage of a clinical trial.

    A   to eliminate irrational preference or prejudice

    B   to ensure that the recipients of the treatment's active factor remain unknown to both the doctors and the subjects

    C   to assess the effects of receiving the treatment that are not dependent on receiving the active ingredient

    D   to reduce experimental error to a minimum

4   Why is a very large sample population of people used in the trial?

5   What is the reason for randomisation when forming the test group and the control group?

6   What is the reason for making the trial double-blind?

7   Figure 24.4 represents the results of three phase III trials on a new drug (M) to give relief to sufferers of migraine.

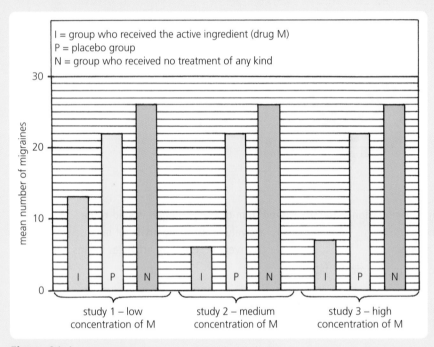

Figure 24.4

Which of the following statements is correct?

    A   The low concentration of drug M brought about a reduction in migraine of 7 units.

    B   The placebo effect is calculated by subtracting the result for I from the result for N.

    C   The efficacy of M increases as its concentration increases.

    D   The efficacy of the drug's active ingredient is measured by subtracting the result for I from the result for P.

8   Herd immunity is a form of protection against a disease given

    A   directly to the non-immune minority by the immune majority.

    B   directly to the non-immune majority by the immune minority.

    C   indirectly to the non-immune minority by the immune majority.

    D   indirectly to the non-immune majority by the immune minority.

9   Which of the following are BOTH factors that may prevent herd immunity to a disease such as poliomyelitis becoming established in a developing country?

    A   poverty and malnutrition   **B**   malnutrition and Rhesus blood group

    C   prevalence of malaria and lack of clean water   **D**   Rhesus blood group and rejection of vaccine

10 The percentage of immune individuals in a population above which a disease no longer manages to persist is called the herd immunity threshold.

It does NOT depend on

A the pathogen's virulence.

B the population's contact parameters.

C the vaccine's efficacy.

D the dominant blood group present in the population.

Questions 11 and 12 refer to Figure 24.5, which shows the death rate of children due to measles in England and Wales.

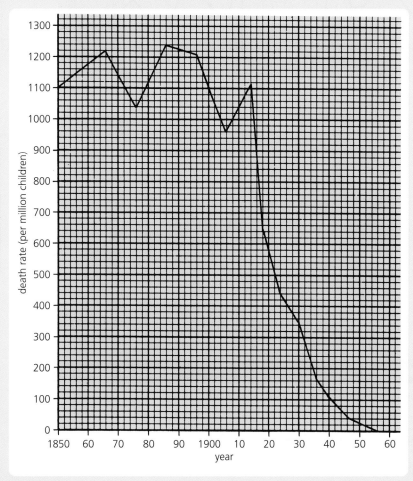

Figure 24.5

11 By how many times was the death rate greater in 1890 than in 1930?

A 2.5   B 3.0   C 3.5   D 4.0

12 Compared with the death rate in 1900, what percentage reduction had occurred by 1940?

A 10   B 90   C 99   D 110

13 Which of the graphs in Figure 24.6 best represents the eradication of smallpox?

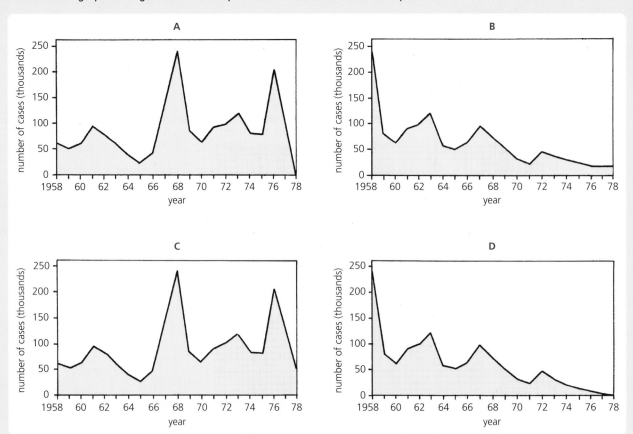

Figure 24.6

Questions 14 and 15 refer to Table 24.1, which shows the immunisation schedule recommended for children in Britain.

Age	Vaccine	Number of injections
2 months	diphtheria + tetanus + pertussis + polio + hib	1
	pneumococcal conjugate	1
3 months	diphtheria + tetanus + pertussis + polio + hib	1
	meningitis C	1
4 months	diphtheria + tetanus + pertussis + polio + hib	1
	pneumococcal conjugate	1
	meningitis C	1
12–13 months	measles + mumps + rubella	1
	pneumococcal conjugate	1
	meningitis C + hib	1
3–5 years	diphtheria + tetanus + pertussis + polio	1
	measles + mumps + rubella	1
12–13 years (girls only)	human papillomavirus (HPV)	3
13–18 years	diphtheria + tetanus + polio	1

Table 24.1

14 The number of doses of pneumococcal conjugate vaccine received by a person who has completed the schedule would be

   A 2   **B** 3   **C** 4   **D** 5

15 A child who has completed the full schedule will have received a total of five doses of vaccine for

   A diphtheria, polio and pertussis.   **B** pertussis, tetanus and polio.

   C polio, diphtheria and tetanus.   **D** tetanus, pertussis and diphtheria.

16 Which of the following statements is NOT correct?

   A Sufferers of AIDS need to be immunised with a new version of vaccine every year to give protection against new strains of the virus.

   B Each new strain of the pathogen that causes sleeping sickness has antigens on its surface that differ from those on the previous strain.

   C Antigenic variation enables new strains of the malarial pathogen to avoid the effects of the host's immunological memory.

   D Each new strain of influenza virus is genetically and immunologically distinct from its parent strain.

17 The flow diagram in Figure 24.7 refers to infection of a person by the influenza virus.

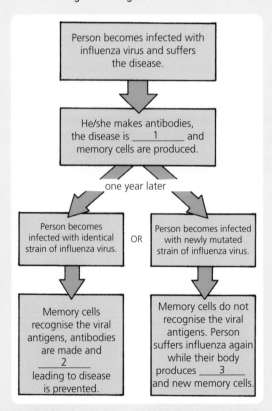

Figure 24.7

Which row in Table 24.2 gives the correct answers to blanks 1, 2 and 3 in Figure 24.7?

	Blank 1	Blank 2	Blank 3
A	contained	antigenic variation	antigens
B	postponed	antigenic variation	antibodies
C	contained	reinfection	antibodies
D	postponed	reinfection	antigens

Table 24.2

18 Which of the following statements is correct? Human immunodeficiency virus (HIV) attacks
    A lymphocytes and is responsible for acquired immune deficiency syndrome.
    B phagocytes and is responsible for antigenic variation.
    C lymphocytes and is responsible for antigenic variation.
    D phagocytes and is responsible for acquired immune deficiency syndrome.

19 Following the invasion of a host cell by HIV, reverse transcriptase present in the HIV promotes the formation of
    A viral RNA from viral DNA.   **B** viral RNA from host DNA.
    C viral DNA from viral RNA.   **D** viral DNA from host RNA.

20 Which of the following statements relating to HIV and AIDS is NOT correct?
    A B lymphocytes make antibodies in response to HIV.
    B AIDS leaves the human body susceptible to opportunistic infections.
    C HIV mutates frequently, forming new strains with different antigenic properties.
    D Some drugs cure AIDS by inhibiting the action of reverse transcriptase.

21 During normal phagocytosis the following stages occur.
    A A macrophage engulfs a bacterial pathogen.
    B The bacterium becomes enclosed in a vesicle.
    C Lysosomes fuse with the vesicle containing the bacterium.
    D Digestive enzymes destroy the pathogen.

    Which of these stages is prevented by molecules released from the cell wall of the bacterium responsible for tuberculosis (TB), enabling it to avoid detection by the immune system?

Questions 22, 23 and 24 refer to the data in Table 24.3, which refer to a worldwide infectious disease.

World Health Organization (WHO) region	Total number suffering disease at start of 2008 ($\times 10^4$)	Number of new cases diagnosed during 2008 ($\times 10^4$)	Number of deaths from the disease during 2008 ($\times 10^4$)
The Americas	22	28	3
Europe	30	42	5
Eastern Mediterranean	85	67	11
Western Pacific	198	194	26
South-East Asia	380	320	47
Africa	385	282	38
World total	1100	933	130

Table 24.3

22 At the start of 2008, which WHO region suffered 18% of the world's total number of cases of this disease?
    A Eastern Mediterranean   **B** Europe   **C** South-East Asia   **D** Western Pacific

23 At the start of 2008, the total number of people suffering the disease in Africa was greater than that in the Americas by a factor of
    A 17.5 times.   **B** 36.3 times.   **C** 175 times.   **D** 3630 times.

24 The total number of people in Europe and the Eastern Mediterranean suffering the disease at the end of 2008 was
    A 198×10^4 **B** 208×10^4 **C** 218×10^4 **D** 240×10^4

25 Typhoid and cholera were still widespread in Britain until 1875 when the Public Health Act was passed. From that time on, the incidence of the diseases steadily declined. Which of the following factors BOTH contributed largely to the EARLY decline in the incidence of these diseases?
    A clean water and improved nutrition   **B** improved nutrition and vaccination
    C vaccination and antibiotics   **D** antibiotics and clean water

Specimen Examination 1

Choose the ONE correct answer to each of the following multiple choice questions.

1 The following list gives the steps that may be used in the future to culture and make use of human stem cells.

   1 Stem cells cloned into colonies in the laboratory.
   2 Differentiated cells used to repair damaged organs.
   3 Stem cells induced by chemical means to differentiate.
   4 Undifferentiated cells extracted from an embryo.

Which of the following is the correct sequence of steps?

   A 4, 1, 3, 2   B 4, 1, 2, 3   C 1, 4, 3, 2   D 1, 4, 2, 3

2 Figure 25.1 represents the human life cycle.

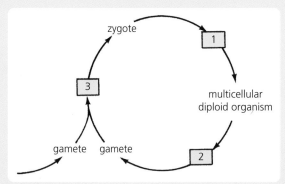

Figure 25.1

Which of the following combinations of terms correctly matches the numbered boxes?

	Box 1	Box 2	Box 3
A	mitosis	fertilisation	meiosis
B	meiosis	mitosis	fertilisation
C	mitosis	meiosis	fertilisation
D	fertilisation	meiosis	mitosis

Table 25.1

3 Which row in Table 25.2 represents DNA?

	Relative percentage of base				
	adenine	guanine	uracil	cytosine	thymine
A	23	27	0	27	23
B	23	27	0	23	27
C	29	21	24	26	0
D	29	29	0	21	21

Table 25.2

4   Figure 25.2 shows a molecule of tRNA.

Figure 25.2

The region labelled X is called

A   a codon. B   an anticodon.

C   a regulator gene. D   an amino acid attachment site.

5   Figure 25.3 shows two chromosomes. The lettered regions represent genes.

Figure 25.3

Which of the following would result if a translocation occurred between chromosomes 1 and 2?

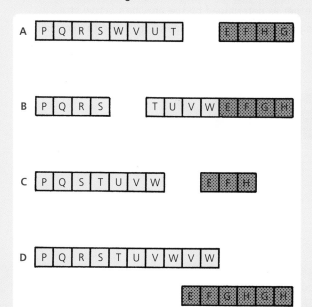

Figure 25.4

6 Figure 25.5 shows several sets of genetic fingerprints used in an attempt to solve a crime. Which suspect's DNA was present at the scene of the crime?

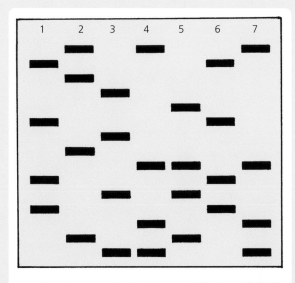

key
1 = DNA sample from victim of crime
2 = DNA sample from suspect **A**
3 = DNA sample from suspect **B**
4 = DNA sample from suspect **C**
5 = DNA sample from suspect **D**
6 = first sample from forensic evidence
7 = second sample from forensic evidence

Figure 25.5

7   The following list gives descriptions of three aspects of the relationship between an enzyme and its substrate(s).
    P = way in which molecules of two reactants are held together as determined by the enzyme's active site
    Q = state of close molecular contact resulting from a change in shape of the enzyme's active site to accommodate its substrate
    R = complementary relationship between a molecule of enzyme and its substrate
    Which row in Table 25.3 correctly matches these aspects with their descriptions?

	Aspect of enzyme-substrate relationship		
	induced fit	specificity	orientation
A	Q	P	R
B	P	R	Q
C	R	Q	P
D	Q	R	P

Table 25.3

8   Which of the following is the correct sequence of the processes that occur during cellular respiration?
    A   glycolysis ⟶ citric acid cycle ⟶ electron transport chain
    B   citric acid cycle ⟶ glycolysis ⟶ electron transport chain
    C   glycolysis ⟶ electron transport chain ⟶ citric acid cycle
    D   electron transport chain ⟶ citric acid cycle ⟶ glycolysis chain

9   Figure 25.6 shows the male reproductive system. Which structure contains seminiferous tubules?

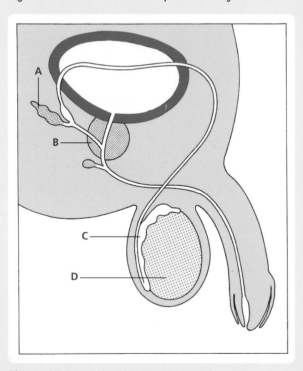

Figure 25.6

10 Which of the following hormones is responsible for triggering ovulation?
    A   follicle-stimulating hormone   B   luteinising hormone
    C   progesterone   D   oestrogen

11 Figure 25.7 shows a simplified version of a biochemical pathway that normally occurs during cell metabolism in humans. Which gene has undergone a mutation in a sufferer of phenylketonuria?

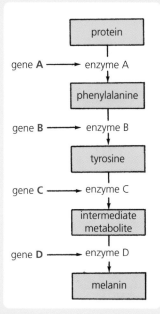

Figure 25.7

12 Atherosclerosis is the formation of atheromas
   A in the lumen of an artery.
   B in the connective tissue in the wall of a vein.
   C beneath the endothelium in the wall of an artery.
   D beneath the thin muscular layer in the wall of a vein.

13 Which row in Table 25.4 indicates four factors that are ALL associated with increased risk of CVD (cardiovascular disease)?

	Smoking	High HDL-cholesterol level	High blood pressure	Physical inactivity	Diet high in unsaturated fats	High LDL-cholesterol level
A	✓		✓	✓		✓
B	✓	✓		✓	✓	
C	✓		✓		✓	✓
D		✓	✓	✓	✓	

Table 25.4

14 Following increased secretion of epinephrine, dilation (increase in width of bore) would occur in the arterioles supplying blood to the
   A skin.
   B skeletal muscles.
   C small intestine.
   D kidneys.

15 Figure 25.8 shows the internal structure of part of the brain.

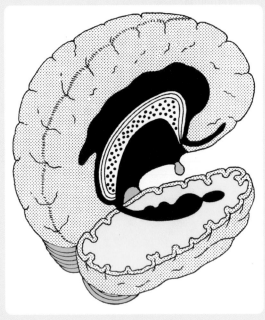

Figure 25.8

The region shaded in black is called the

A medulla. B limbic system. C corpus callosum. D cerebral hemisphere.

16 Figure 25.9 shows the action of a normal natural neurotransmitter in the absence of drugs. Which part of Figure 25.10 on page 175 correctly shows an agonist drug in action?

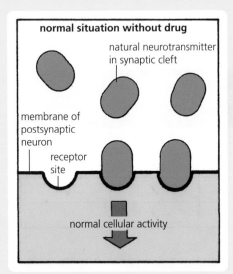

Figure 25.9

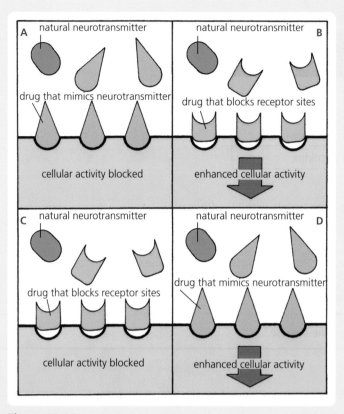

Figure 25.10

17 Figure 25.11 shows two surrogate mothers used in an experiment to investigate the effect of bodily contact on infant monkeys.

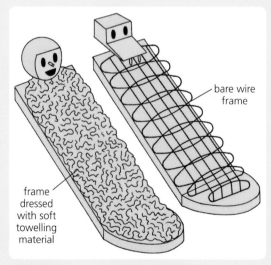

bare wire frame

frame dressed with soft towelling material

Figure 25.11

This experiment would be valid if

A the surrogate mothers' heads were exactly the same.

B a live mother monkey was used as a control.

C both mothers were covered in soft towelling material.

D only the wire mother was used to provide the infants with food.

18 Table 25.5 shows three cell types and chemicals that they release. Which row is correct?

	Cell type		
	natural killer cell	**mast cell**	**phagocyte**
A	cytokine	cytokine	histamine
B	histamine	cytokine	lysozyme
C	cytokine	histamine	cytokine
D	lysozyme	histamine	cytokine

Table 25.5

19 A person with blood group O can safely donate blood to people with blood group(s)

A O only.
                                 B A and B only.

C A, B and AB only.
                   D A, B, AB and O.

20 The length of the quarantine period for a person suspected of having an infectious disease is normally

A determined by the age of the infected person.

B determined by the country in which the disease occurs.

C equal to the maximum known incubation period of the disease.

D equal to the maximum known duration period of the disease.

Specimen Examination 2

Choose the ONE correct answer to each of the following multiple choice questions.

1 Which of the following types of tissue is characterised by having a large quantity of extracellular material present in the spaces between its cells?

A connective   B epithelial   C muscle   D nerve

2 Figure 26.1 shows the growth of a culture of cancer cells in 70 cm^3 of liquid medium over a period of 60 hours.

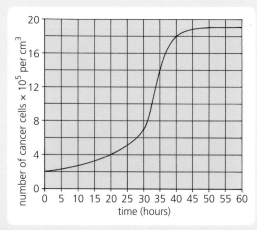

Figure 26.1

The total number of cells present at 30 hours was

A 4.90×10^6 B 5.25×10^6 C 4.90×10^7 D 5.25×10^7

3 The set of results in Table 26.1 shows an analysis of the DNA bases contained in the cells of a cow's thymus gland.

Base composition (%)			
X	guanine	Y	Z
28.7	21.5	21.3	28.4

Table 26.1

Which of the following is a possible correct identification of the bases?

	X	Y	Z
A	adenine	cytosine	thymine
B	thymine	adenine	cytosine
C	cytosine	adenine	thymine
D	cytosine	thymine	adenine

Table 26.2

4   Figure 26.2 shows a molecule of DNA before replication has occurred.

If ——— represents an original DNA strand and – – – – – represents a new DNA strand, which of the daughter molecules in Figure 26.3 will result from the replication of the DNA molecule shown in Figure 26.2?

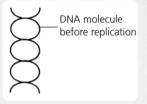

DNA molecule before replication

Figure 26.2

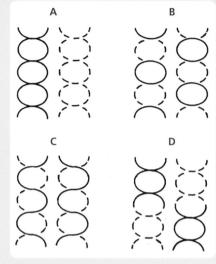

A   B

C   D

Figure 26.3

5   The flow chart in Figure 26.4 refers to the coding for and synthesis of an active protein. Which arrow represents the process of translation?

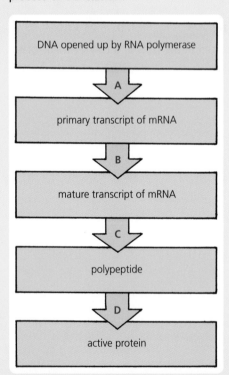

DNA opened up by RNA polymerase

A

primary transcript of mRNA

B

mature transcript of mRNA

C

polypeptide

D

active protein

Figure 26.4

6   What name is given to the type of point mutation where one incorrect nucleotide occurs in place of the correct nucleotide in a DNA chain?

A   deletion   B   insertion   C   inversion   D   substitution

7 Which row in Table 26.3 correctly identifies blanks 1, 2 and 3 in the following sentence?
   The analysis of personal _____1_____ may lead to _____2_____ medical treatment based on
   _____3_____ genomes in the future.

	Blank 1	Blank 2	Blank 3
A	bioinformatics	standardised	individual
B	genomes	standardised	typical
C	bioinformatics	personalised	typical
D	genomes	personalised	individual

Table 26.3

8 Figure 26.5 shows a metabolic pathway that can be controlled by end product inhibition.

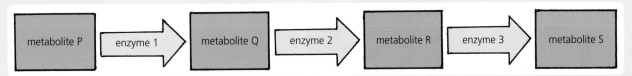

Figure 26.5

Metabolite S would bring about this process of end product inhibition by interacting with
A metabolite P.   B enzyme 1.   C metabolite R.   D enzyme 3.

9 Figure 26.6 shows changes that take place in the endometrium during a normal menstrual cycle. Which arrow
   represents ovulation?

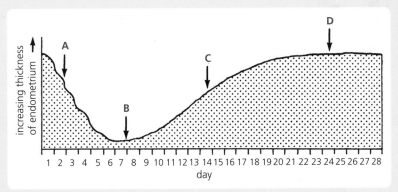

Figure 26.6

10 Table 26.4 refers to four samples of semen. Which sample contains the lowest number of active sperm?

Semen sample	Number of sperm in sample (millions/cm³)	Active sperm (%)
A	20	80
B	30	50
C	40	60
D	50	40

Table 26.4

11 Which of the following is NOT a procedural step carried out during amniocentesis?

A sample of placental cells taken using a catheter

B cells cultured and examined under a microscope

C chromosomes photographed and arranged in pairs

D fluid containing fetal cells withdrawn using a syringe

12 The bar chart in Figure 26.7 shows the rate of blood flow in various parts of a person's body under differing conditions of exercise.

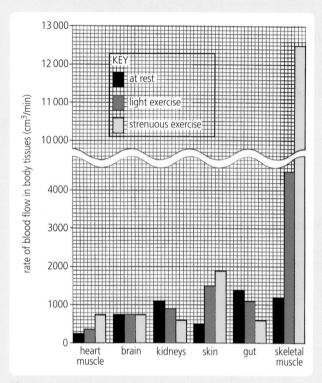

Figure 26.7

Which body parts ALL show the same trend in response to increased exercise?

A skin, gut and heart muscle

B gut, kidneys and brain

C brain, heart muscle and kidneys

D skeletal muscle, skin and heart muscle

13 The statements listed in Figure 26.8 give some of the events that lead to a pulmonary embolism.

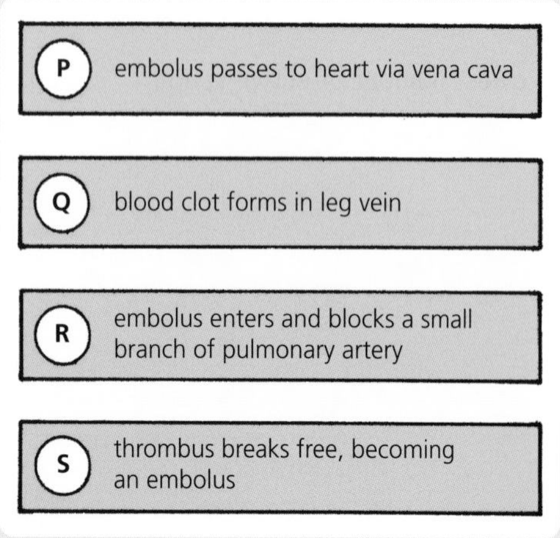

P embolus passes to heart via vena cava

Q blood clot forms in leg vein

R embolus enters and blocks a small branch of pulmonary artery

S thrombus breaks free, becoming an embolus

Figure 26.8

The order in which these occur is

A Q, S, R, P. B Q, S, P, R. C S, Q, R, P. D S, Q, P, R.

14 Which of the following numbered statements are BOTH correct?

1 Glucagon activates the conversion of glucose to glycogen, thereby lowering the blood sugar level.
2 Glucagon activates the conversion of glycogen to glucose, thereby raising the blood sugar level.
3 Insulin activates the conversion of glycogen to glucose, thereby raising the blood sugar level.
4 Insulin activates the conversion of glucose to glycogen, thereby lowering the blood sugar level.

A 1 and 2 B 1 and 4 C 2 and 3 D 2 and 4

15 Figure 26.9 shows the left cerebral hemisphere. Which region (on receiving messages from other parts of the brain) sends motor impulses to skeletal muscles, bringing about movement?

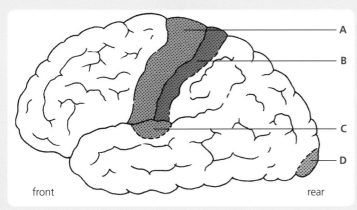

Figure 26.9

16 A possible reason for having difficulty retrieving an old memory that was once distinct is that the

A item was not successfully encoded in the first place.
B necessary contextual cues are missing.
C information was over-rehearsed at the time.
D memory is spread over too many categories.

17 Which of the diagrams in Figure 26.10 best represents the arrangement of the three types of neuron found in a reflex arc?

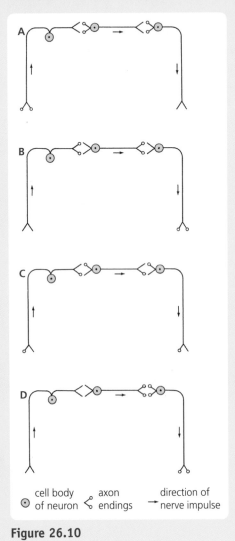

Figure 26.10

18 The type of learning behaviour shown in Figure 26.11 is called

   A imitation.   **B** motivation.   **C** reinforcement.   **D** trial and error.

Figure 26.11

19 Which of the following are BOTH features characteristic of the inflammatory response?

   A vasoconstriction of arterioles and increased capillary permeability

   B increased capillary permeability and antibody production

   C antibody production and release of histamine

   D release of histamine and vasodilation of arterioles

20 Which of the following statements is correct?

   A An antibody stimulates the body to produce antigens.

   B An antigen stimulates white blood cells to make antibodies.

   C An antibody is always composed of toxin from a pathogen.

   D An antigen is always composed of viral protein.

Answer grid

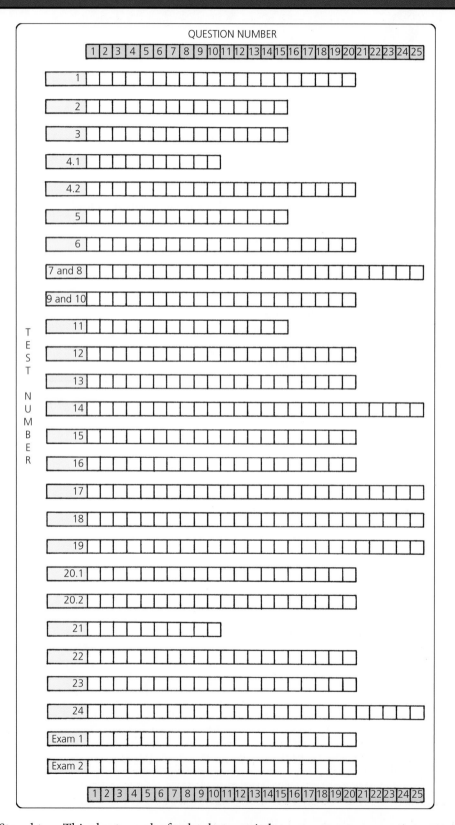

Completed answer grid

QUESTION NUMBER

Test No.	1	2	3	4	5	6	7	8	9	10	11	12	13	14	15	16	17	18	19	20	21	22	23	24	25
1	D	C	C	A	C	A	B	B	D	D	B	C	A	D	A	C	A	B	D	B					
2	A	C	A	B	B	D	C	C	D	D	B	A	B	B	C										
3	A	C	B	D	A	A	B	C	D	A	D	B	C	C	D										
4.1	A	C	C	A	C	B	A	B	D	D															
4.2	D	D	B	D	A	B	A	B	D	C	A	A	B	B	D	C	C	A	B	C					
5	D	A	C	A	C	B	D	B	A	C	C	D	B	D	A										
6	D	B	A	C	B	A	A	C	D	D	A	B	C	C	D	A	D	B	C	B					
7 and 8	B	A	B	C	C	D	A	C	D	B	C	D	A	D	B	C	A	B	A	B	C	D	A	D	B
9 and 10	A	B	D	D	C	D	B	D	A	B	A	A	C	B	A	B	C	C	D	C					
11	B	A	C	C	D	C	B	A	A	A	D	B	B	D	C										
12	B	D	A	A	D	C	A	B	A	C	B	A	B	D	C	D	C	C	B	D					
13	A	B	D	C	B	B	D	A	A	B	C	D	B	C	D	D	A	A	C	C					
14	A	C	D	B	B	A	C	D	A	C	B	D	A	C	C	D	A	B	D	B	A	C	B	D	B
15	D	A	D	C	C	B	C	B	A	D	A	C	D	B	A	C	B	A	B	D					
16	A	D	A	D	C	B	B	D	D	C	A	C	B	C	B	D	B	A	C	B					
17	D	A	C	D	C	B	C	A	D	D	A	D	A	C	B	B	D	B	C	B	A	C	A	D	B
18	B	A	D	B	A	C	C	D	A	D	C	C	A	C	B	C	A	D	A	B	D	B	B	A	D
19	B	C	A	D	C	D	B	A	C	B	D	A	A	C	D	B	D	B	A	C	D	B	C	C	A
20.1	C	D	B	A	D	C	D	C	B	A	D	C	B	D	A	A	B	B	C	A					
20.2	A	B	C	D	D	A	C	B	A	D	B	C	A	D	B	C	B	C	D	A					
21	C	D	D	B	D	A	A	B	D	A															
22	D	B	A	A	C	D	C	C	A	D	B	B	A	D	B	B	B	D	C	A	D				
23	D	C	B	A	D	C	C	A	D	B	D	A	D	C	A	B	C	A	B	B					
24	A	B	B	D	A	B	D	C	A	D	C	B	D	B	C	A	C	A	C	D	C	D	A	B	A
Exam 1	A	C	A	D	B	C	D	A	D	B	B	C	A	B	B	D	A	C	D	C					
Exam 2	A	C	A	C	C	D	D	B	C	B	A	D	B	D	A	B	B	A	D	B					

QUESTION NUMBER 1 2 3 4 5 6 7 8 9 10 11 12 13 14 15 16 17 18 19 20 21 22 23 24 25